Urban Air Pollution in Asian Cities

Status, Challenges and Management

Dieter Schwela, Gary Haq, Cornie Huizenga,
Wha-Jin Han, Herbert Fabian and May Ajero

Routledge
Taylor & Francis Group

LONDON AND NEW YORK

First published 2006 by Earthscan

2 Park Square, Milton Park, Abingdon, Oxon OX14 4RN
711 Third Avenue, New York, NY 10017, USA

Routledge is an imprint of the Taylor & Francis Group, an informa business

First issued in paperback 2016

ISBN-13: 978-1-138-98658-9 (pbk)

Typesetting by JS Typesetting Ltd, Porthcawl, Mid Glamorgan
Cover design by Andrew Corbett

*The contents of this book form Stage II of an exercise to benchmark urban air quality management in 20 Asian
cities.*

*This publication was produced as part of the Air Pollution in the Megacities of Asia (APMA) project in collaboration
with the Clean Air Initiative for Asian Cities (CAI-Asia).*

*The contents of this book do not necessarily reflect the views or policies of the Asian Development Bank, CAI-
Asia including its members, Korea Environment Institute, Korean Ministry of Environment, Stockholm Environment
Institute, Swedish International Development Cooperation Agency, World Health Organization and the United
Nations Environment Programme.*

A catalogue record for this book is available from the British Library

Library of Congress Cataloging-in-Publication Data
Urban air pollution in Asian cities : status, challenges and management /
Dieter Schwela ... [et al.].
    p. cm.
  ISBN-13: 978-1-84407-375-7 (hardback)
  ISBN-10: 1-84407-375-9 (hardback)
  1. Air–Pollution–Asia. 2. Metropolitan areas–Asia. 3. Urban
pollution–Asia. 4. Air quality management–Asia. I. Schwela, Dietrich.
  TD883.7.A78U72 2006
  363.739'2095091732–dc22
                                       2006014538

In memory of our colleague and friend
Kong Ha
31 July 1955–3 April 2007
Chair of the Clean Air Initiative for Asian Cities 2004–2007

Contents

List of Tables and Figures

TABLES

FIGURES

Acknowledgements

We are grateful to a large number of individuals who have contributed to various stages of the study. This work would not have been completed without the assistance of international and regional air pollution experts and representatives from the 20 cities who provided information and reviewed our work. We are indebted to all of them:

Bangkok

Supat Wangwongwatana, Deputy Director General, Pollution Control Department, Thailand

Mingquan Wichayarangsaridh and staff, Air Quality and Noise Management Bureau, Pollution Control Department, Thailand

Beijing

Yu Tong, Deputy Director, Beijing Environmental Monitoring Center, Beijing, China

Busan

OkTaek Kim, Director, Enviromental Conservation Division, Busan, Republic of Korea

Colombo

C. K. Amaratunge, Air Management Resources Center, Colombo, Sri Lanka
Don Jayaweera, Ministry of Transport, Colombo, Sri Lanka

Ananda Mallawatantri, US-AEP, Colombo, Sri Lanka

Ruwan Weerasooriya, Research Assistant, Air Management Resources Center, Colombo, Sri Lanka

Dhaka

Shahana Akhter, Deputy Director, Department of Environment, Dhaka, Bangladesh

John Core, Core Environmental Consulting, Portland, Oregon, USA

Derek Langgons, Project Consultancy Coordinator, Air Quality Management Project, Dhaka, Bangladesh

Hanoi

Dao Xuan Duc, Program Officer, Swiss–Vietnamese Clean Air Program (SVCAP), Swisscontact, Hanoi, Vietnam

Nguyen Hoai Anh, Specialist, Vietnam Register, Hanoi, Vietnam

Le Tran Lam, Director, Hanoi-DOSTE, Hanoi, Vietnam

Phan Quynh Nhu, Clean Cities Coordinator, Swisscontact, (Previously US-AEP/TSSC), Hanoi, Vietnam

Ho Chi Minh City

Vo Than Dam, Head of Environmental Quality Monitoring and Assessment Division, Ho Chi Minh City Environmental Protection Agency, Ho Chi Minh City, Vietnam

Nguyen Dinh Tuan, Director, Ho Chi Minh City Environmental Protection Agency, Ho Chi Minh City, Vietnam

Le Van Khoa, Deputy Head, Ho Chi Minh City Environmental Protection Agency, Ho Chi Minh City, Vietnam

Bjarne Sivertsen, Associate Research Director, Norwegian Institute for Air Research (NILU), Oslo, Norway

Hong Kong

Kong Ha, Senior Environmental Protection Officer, Environmental Protection Department, Hong Kong, SAR China

Albert Leung, Environmental Protection Department, Hong Kong, SAR China

Jakarta

Paul Butarbutar, Coordinator, Swisscontact, Jakarta, Indonesia

Esrom Hamonangan, Environment Management Center, Ministry of Environment, Jakarta, Indonesia

Amelia Rachmatunisa, Head, Air Quality Monitoring Sub Division, Ministry of Environment, Jakarta, Indonesia

Ridwan Tamin, Assistant Deputy for Vehicle Emissions Affairs, Ministry of Environment, Jakarta, Indonesia

Kathmandu

Anil Raut, Program Officer, Environmental Pollution, Environment Nepal, Kathmandu, Nepal

Rabin Man Shrestha, Chief, Urban Environment Section, Kathmandu Metropolitan City Office, Kathmandu, Nepal

Kolkata

Prasenjit Bhattacharya, Scientist, West Bengal Pollution Control Board, Kolkata, India

Metro Manila

Cesar Siador, Chief, Air Quality Management Section, Environmental Management Bureau, Department of Environment and Natural Resources, Metro Manila, Philippines

Genie Lorenzo, Urban Air Quality Group, Manila Observatory, Metro Manila, Philippines

Tess Peralta, Air Quality Management Section, Environmental Management Bureau, Department of Environment and Natural Resources, Metro Manila, Philippines

Jean Rosete, Deputy Program Director, Air Quality Management Section, Environmental Management Bureau, Department of Environment and Natural Resources, Metro Manila, Philippines

Mumbai

Debi Goenka, Bombay Environmental Action Group, Mumbai, India

Rakesh Kumar, Senior Assistant Director, National Environmental Engineering Research Institute (NEERI), Mumbai, India

Anjali Srivastava, Senior Assistant Director, National Environmental Engineering Research Institute (NEERI), Mumbai, India

New Delhi

Naresh Badhwar, Assistant Environmental Engineer, Central Pollution Control Board, New Delhi, India

Anumita Roychowdhury, Associate Director, Research and Advocacy, Centre for Science and Environment, New Delhi, India

B. Sengupta, Member Secretary, Central Pollution Control Board, New Delhi, India

Seoul

Hee-Jeong Chae, Deputy Director, Air Quality Control Division, Environmental Bureau, Seoul, Republic of Korea

Shanghai

Jacqueline (Qingyan) Fu, Assistant Director, Shanghai Environmental Monitoring Centre, Shanghai, China

Singapore

Joseph Foo, National Environment Agency, Planning and Development Department, Singapore

Joseph Hui, Head, Planning and Development Department, National Environment Agency, Singapore

Soh Suat Hoon, Senior Scientific Officer, National Environment Agency, Planning and Development Department, Singapore

Surabaya

Chamidha, Head of Air Laboratory, Surabaya City Environmental Management Agency (BPLHD), Surabaya, Indonesia

Driejana, Air Laboratory Researcher, Faculty of Civil and Environmental Engineering, Institut Teknologi Bandung, Surabaya, Indonesia

Togar Arifin Silaban, Chief, Division of Urban Infrastructure, Surabaya Development Planning Agency, Surabaya, Indonesia

Taipei

Li-Ling Huang, Chief, Air Pollution Management Division, Bureau of Environmental Protection, Taipei Municipal Government, Taipei, China

Gloria Kung Jung-Hsu, Vice Chair, Taiwan Environmental Protection Union, Taipei, China

Tokyo

Yuko Nishida, Bureau of Environment, Tokyo Metropolitan Government, Tokyo, Japan

Yoshihito Takeda, General Affairs Division, Bureau of Environment, Tokyo Metropolitan Government, Tokyo, Japan

We would like to thank Hiremagular N. B. Gopalan, Jae-Hyun Lee and Youn Lee of UNEP and the Ministry of Environment Korea for initiating and supporting the APMA project. We would also like to thank Christine Kim for all her hard work in setting up the study. It was great fun working with her. Thank you also to John E. Hay, Harry Vallack and Michael J. Chadwick for their contribution to Stage I.

This study was funded by the Korean Ministry of Environment, Swedish International Development Agency, the Clean Air Initiative for Asian Cities and the Asian Development Bank with in-kind contributions from WHO and UNEP. We are grateful to these agencies for also providing the resources to undertake the research for this book.

We would like to thank the following international reviewers for their helpful comments and suggestions:

Mike Ashmore, Department of Environment, University of York, UK

Bingheng Chen, School of Public Health, Fudan University, Shanghai, China

Lidia Morawska, Queensland University of Technology, Brisbane, Queensland, Australia

Frank Murray, School of Environmental Science, Murdoch University, Perth, Australia

Michal Krzyzanowski, World Health Organization, Regional Office for Europe, Bonn, Germany

Bjarne Sivertsen, Norwegian Institute for Air Research (NILU), Kjeller, Norway

At SEI, we would like to thank Bingyan (Isabel) Wang for her contribution to the finalization of the study, Lisetta Tripodi and Jenny Duckmanton for proof-reading. In particular we would like to acknowledge the sterling work of Erik Willis in preparing the figures. We thank him for his patience and chocolate biscuits!

Finally, we would like to thank Rob West and colleagues at Earthscan for their assistance in the preparation of this book.

Preface

Air pollution is part of the daily existence of the many thousands of people who work, live and use the streets in Asian cities. The poor and socially marginalized in particular tend to suffer disproportionately from the effects of deteriorating air quality. Despite improvements in many Asian cities, air pollution is still relatively high compared to cities in the developed world. Particulate matter, nitrogen dioxide and ozone are key pollutants that still pose a significant challenge to improving urban air quality.

There is a growing need to determine the state of urban air quality and to identify the most effective measures to protect human health and environment. Learning from experience and successes in urban air quality management from other countries can assist in the formulation and implementation of strategies to achieve better air quality in Asia.

This book provides the results of an exercise to benchmark air quality management in Asian cities initiated by the Air Pollution in the Megacities of Asia (APMA) project, which involved the Stockholm Environment Institute (SEI), the Korea Environment Institute (KEI), the United Nations Environment Programme (UNEP) and the World Health Organization (WHO). The study was completed in collaboration with the Clean Air Initiative for Asian Cities (CAI-Asia). It builds upon the APMA Benchmarking Stage I Report (2002), which provided an initial assessment of 12 Asian cities.

The aim of the study on which this book is based was to determine the current status of air quality management in 20 Asian cities. It builds upon the UNEP/WHO (1992) *Urban Air Pollution in the Megacities of the World* and the MARC/WHO/ UNEP (1996) *Air Quality Management Assessment Capabilities in 20 Major Cities* reports as well as the WHO (2001) Air Management Information System (AMIS). It is intended to provide a benchmark from which future initiatives and progress in Asia can be assessed and to allow the exchange of lessons learnt in addressing urban air pollution issues in different countries.

The book provides the first comprehensive and systematic assessment of 20 mega- and major cities in Asia. It assesses the current status, challenges and management in urban air pollution in Bangkok, Beijing, Busan, Colombo, Dhaka, Hanoi, Ho Chi Minh City, Hong Kong, Jakarta, Kathmandu, Kolkata, Metro Manila, Mumbai, New Delhi, Seoul, Shanghai, Singapore, Surabaya, Taipei and Tokyo. It determines the air quality management capability of each city and the current trends and challenges they face in managing air quality. It identifies the stage of air pollution development for each city, which can assist cities in setting priorities and developing strategies to strengthen their air quality management capability. By monitoring progress, learning from experience and identifying priorities, national and local authorities can be assisted in achieving a better quality of life for the millions of people living in Asian cities.

Despite gaps in data, which are reflective of the challenges of air quality management in the cities, and limitations in methodology used, we believe this book will provide a useful resource on the current status of urban air quality in Asia. It is a timely contribution to the exchange of information and knowledge on management of air quality and will be of interest to policy makers, air pollution practitioners, students and anyone else interested in achieving clean air in Asian cities.

REFERENCES

Haq, G., Han, W., Kim, C. and Vallack, H. (2002) 'Benchmarking urban air quality management and practice in major and minor cities in Asia Stage I', Air Pollution in the Megacities of Asia, www.cleanairnet.org/caiasia/1412/propertyvalue-15186. html accessed in May 2006

UNEP/WHO (1992) *Urban Air Pollution in Megacities of the World*. United Nations Environment Programme/World Health Organization, Oxford, Blackwell

WHO (2001) 'Air Quality and Health, Air Management Information System (AMIS 3.0)', (CD-ROM), World Health Organization, Geneva

WHO/UNEP/MARC (1996) *Air Quality Management and Assessment Capabilities in 20 Major Cities*, UNEP/DEIA/AR.96.2/WHO/EOS.95.7, United Nations Environment Programme, Nairobi, Kenya

List of Acronyms and Abbreviations

ABC	atmospheric brown cloud
ADB	Asian Development Bank
AEAT	AEA Technology Environment
AirMAC	Air Resource Management Centre (Sri Lanka)
AMIS	Air Management Information System
APCA	Air Pollution Control Act
APCEL	Asia-Pacific Centre for Environmental Law
APCO	Air Pollution Control Ordinance
API	Air Pollution Index
APMA	Air Pollution in the Megacities of Asia
AQ	air quality
AQM	air quality management
AQMP	Air Quality Management Project
AQMS	air quality management system
AQO	Air Quality Objective
ARI	acute respiratory infection
ARRPET	Asian Regional Research Program on Environmental Technology
ASEAN	Association of Southeast Asian Nations
BaP	Benzo(a)pyrene
BAPEDAL	Indonesian Environmental Impact Management Agency
BAQ	Better Air Quality
BJEPB	Beijing Environmental Protection Bureau
BMA	Bangkok Metropolitan Administration
BMR	Bangkok Metropolitan Region
BP	British Petroleum

BRT	bus rapid transit
CAA	Clean Air Act
CACL	Clean Air Conservation Law
CAFE	Clean Air for Europe
CAI-Asia	Clean Air Initiative for Asian Cities
CEA	Central Environmental Authority
CEC	cation exchange capacity
CEN	Clean Energy Nepal
CMR	Colombo Metropolitan Region
CNG	compressed natural gas
CO	carbon monoxide
CO_2	carbon dioxide
COPD	chronic obstructive pulmonary disease
CPCB	Central Pollution Control Board
CSE	Centre for Science and Environment
DENR	Department of Environment and Natural Resources
DEP	Department of Environmental Protection
DOE	Department of Environment
DOSTE	Department of Science, Technology and Environment
DOTC	Department of Transportation and Communications
DPSIR	Driving Force-Pressure-State-Impact-Response framework
DUTP	Dhaka Urban Transport Project
EANET	Acid Deposition Monitoring Network in East Asia
EC	electric conductivity
EC	elemental carbon
ECR	Environmental Conservation Rules
ED	Emergency Department
EEA	European Environment Agency
EKC	Environmental Kuznets Curve
EPD	Environment Protection Department
ENPHO	Environment and Public Health Organization
EPCA	Environmental Pollution Control Act
EPTRI	Environment Protection and Training Research Institute
ESMAP	Energy Sector Management Assistance Programme
ESPS	Environmental Sector Programme Support
EST	Environmentally Sustainable Transport
EU	European Union
FADA	Federation of Automobile Dealers Associations
FEV_1	forced expiry volume in one second
FVC	forced vital capacity

GDP	gross domestic product
GEF	Global Environment Forum
GHG	greenhouse gases
GNI	gross national income
GONCT	Government of National Capital Territory of Delhi
GVW	gross vehicle weight
HC	hydrocarbons
HCMC	Ho Chi Minh City
HEI	Health Effects Institute
HEPA	Ho Chi Minh Environmental Protection Agency
HTAP	hemispheric transport of pollutants
IAEA	International Atomic Energy Agency
IEA	International Energy Outlook
INDOEX	Indian Ocean Experiment
JAMA	Japan Automobile Manufacturers Association
KEI	Korea Environment Institute
KMA	Kolkata Metropolitan Area
LEP	Law on Environmental Protection
LDO	light diesel oil
LEP	Law on Environmental Protection (Vietnam)
LGC	Local Government Code
LPG	liquid petroleum gas
LRTAP	Long-range Transboundary Air Pollution
LSHS	low sulphur heavy stock
LTO	Land Transportation Office
MARC	Monitoring and Assessment Research Centre
MINAS	Minimum National Standards
MOC	Ministry of Construction (Vietnam)
MOE	Ministry of Environment
MoEF	Ministry of Environment and Forests
MOEST	Ministry of Environment, Science and Technology
MONRE	Ministry of Natural Resources and Environment
MOPE	Ministry of Population and Environment
MPCB	Mumbai Pollution Control Board
MRTH	Ministry of Road Transport and Highways
MVIS	Motor Vehicle Inspection System
NASA	National Aeronautics and Space Administration
NBRO	National Building Research Organization
NEA	National Environment Agency
NEERI	National Environmental Engineering Research Institute

NEQA	National Environmental Quality Act (Thailand)
NMHC	non-methane hydrocarbon
NMVOC	non-methane volatile organic compound
NO_2	nitrogen dioxide
NO_x	nitrogen oxides ($NO + NO_2$)
NORAD	Norwegian Agency for Development Cooperation
NRTEE	National Round Table on Environment and the Economy
O_3	ozone
OC	organic carbon
OECD	Organisation for Economic Co-operation and Development
PAH	polycyclic aromatic hydrocarbons
PAPA	Public Health and Air Pollution in Asia
Pb	lead
PCBs	polychlorinated biphenyls
PCD	Pollution Control Department
PCFV	Partnership for Clean Fuels and Vehicles
PETC	Private Emission Testing Center
PM	particulate matter
PPP	power purchasing parity
PRD	Pearl River Delta (Hong Kong)
PSI	Pollutant Standard Index
QA/QC	quality assurance and quality control
RAPIDC	Regional Air Pollution in Developing Countries Programme
RHAP	Regional Haze Action Plan
RSPM	respirable suspended particulate matter
SAR	Special Administrative Region
SARS	severe acute respiratory syndrome
SEI	Stockholm Environment Institute
SEMC	Shanghai Environmental Monitoring Centre
SEPA	State Environmental Protection Agency
Sida	Swedish International Development Cooperation Agency
SO_2	sulphur dioxide
SPCB	State Pollution Control Board
TAQMN	Taiwan Air Quality Monitoring Network
TLEV	Transitional Low Emission Vehicle
TSP	total suspended particulate
UIC	Uranium Information Centre
UK	United Kingdom
ULEV	Ultra Low Emission Vehicle
UN	United Nations

UNCED	United Nations Conference on Environment and Development
UNCRD	United Nations Centre for Regional Development
UNECE	United Nations Economic Commission for Europe
UNEP	United Nations Environment Programme
UNESCAP	United Nations Economic Social Commission for Asia and the Pacific
UNESCO	United Nations Education, Scientific and Cultural Organization
UN-Habitat	United Nations Human Settlements Programme
UNPD	United Nations Population Division
US	United States
USEPA	United States Environmental Protection Agency
USOMB	United States Office of Management and Budget
VOC	volatile organic compound
WBPCB	West Bengal Pollution Control Board
WBSCD	World Business Council for Sustainable Development
WHO	World Health Organization
WMO	World Meteorological Organization
WSSD	World Summit on Sustainable Development

one

Urban Air Pollution in Asia

INTRODUCTION

Urban air pollution affects the health, well-being and life chances of hundreds of millions of men, women and children in Asia every day. It is responsible for an estimated 537,000 premature deaths annually with indoor air pollution being responsible for over double this number of deaths (WHO, 2002). It is often the poor and socially marginalized who tend to suffer disproportionately from the effects of deteriorating air quality due to living near sources of pollution (Martins et al, 2004; Gouveia and Fletcher, 2000; Stern, 2003). The ubiquitous Asian street hawker who sits beside strategic road junctions experiencing the general hustle and bustle of daily life and traffic is being exposed to high concentrations of motor vehicle pollutants increasing the risk of developing respiratory disease and cancer (Chakraborti, 2003). Children ill with respiratory disease caused by exposure to high concentrations of air pollutants will be children who will not learn very well, will suffer in adult life from low levels of qualifications and skills, which in turn has implications for their quality of life and the economic development of the country as a whole.

Levels of air pollution in Asian cities regularly exceed World Health Organization (WHO) recommended guidelines with smoke and dust particles being double the world average (WHO, 2000a; 2005). The main cause of urban air pollution is the use of fossil fuels (coal, oil and natural gas) in transport, power generation, industry and domestic sectors. In addition, the burning of biomass such as firewood, agricultural and animal waste also contributes to pollution levels. Pollutant emissions have direct and indirect effects (e.g. acidification, eutrophication, ground-level ozone, stratospheric ozone depletion) on air quality with a wide range of impacts on human health, ecosystems, agriculture and materials.

The severity of air pollution problems in Asian cities reflects the level and speed of their economic development and the effectiveness of past air quality management

(AQM) and current efforts. Each city is unique in terms of its economic, physical and social characteristics which influence the spatial and temporal patterns of emission sources and air pollution problems.

To gain an understanding of urban pollution in Asia it is necessary to understand the economic development context and the key factors which influence the emission of air pollutants. This chapter examines the development of air pollution problems and the key drivers and pressures of air pollution which affect the current state of air quality. It discusses the key impacts on human health and environment and provides an overview of the measures taken to address air pollution in the region.

DEVELOPMENT OF AIR POLLUTION PROBLEMS

As cities undergo economic development the environmental risks to human health generally decline both in absolute and relative terms with different types of risks tending to dominate each phase of development, moving from the household to community and global scale (Smith, 1997; McGranahan et al, 2001) (see Figure 1.1).

In poor cities, household sanitation, water quality and fuel quality are usually severe problems. Before rapid industrial development, the main source of air pollution is from domestic and low-scale commercial activities. The concentration of air pollution is generally low and it increases as the population increases. As cities develop and undergo industrial development, urbanization and motorization the community

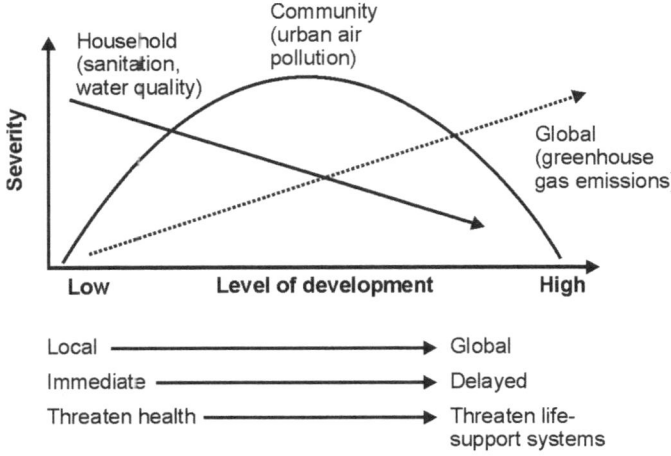

Figure 1.1 *Urban environmental transition model*

Source: Smith (1997); McGranahan et al (2001)

risk from urban activities increases. Outdoor air quality becomes a serious public health issue as carbon emissions increase and the severity of the household sanitation problem decreases. As countries develop further, environmental controls are tightened and community-level risks (urban air pollution) tend to decline. In wealthy cities strict regulation and the implementation of abatement technology reduce polluting emissions to ensure they meet health guidelines, but this is not without considerable financial cost to the community. This decline in community risk leads to the richest countries contributing the most to global risks such as global climate change due to greenhouse gas emissions on a per capita basis. For example, in 2002, the carbon dioxide (CO_2) emissions for North America, Western Europe and Centrally Planned Asia (mostly China) amounted to 1.7, 0.7, and 1.0 trillion metric tons, respectively. However, the absolute CO_2 emissions per capita were 5.42, 2.05 and 0.73 metric tons of carbon respectively, illustrating that the CO_2 emissions from Europe and China together are one-half of the North American emissions (Marland et al, 2005).

The trends in environmental risk at the household, community and global scales across the development process generally support the environmental risk model (Smith and Ezzati, 2005). While there can be significant local departures from this overall trend, the urban environmental transition model provides an aid to understanding the development and management of urban environmental problems and the risk they pose to human health.

The Environmental Kuznets Curve

The 1992 World Bank Development Report noted that 'most forms of pollution ... initially worsen but then improve as incomes rise' (World Bank, 1992). Following this first model, which demonstrated a relationship between air quality and economic growth, a number of studies have further examined the link between air pollution and economic development (Seldon and Song, 1994; Shafik, 1994; Cole et al, 1997; Goklany, 1999; Gangadharan and Valenzuela, 2001; Hill and Magnani, 2002; Cole, 2003). These studies established that the relationship consisted of an inverse U-shape curve, called the Environmental Kuznets Curve (EKC).

The EKC describes the relationship between levels of environmental quality and national economic development. It suggests a certain inevitability of environmental degradation as a country develops, especially during the early stages of industrialization. As the city develops, there is an increase in the use of natural resources, pollutant emissions, use of inefficient and relatively dirty technologies and a high priority is given to material output with a disregard for the environment (Smith, 1997; McGranahan et al, 2001). With increasing energy and resource use, industrialization and motorization, ambient air quality becomes a serious public health problem.

As economic growth continues and life expectancy and income increase, the general public place a greater value on environmental quality and demand better environmental protection (Hettige et al, 1992; Antle and Heidebrink, 1995). Environmental controls are tightened and community-level risks such as urban air pollution tend to decline. In the post-industrial stage, cleaner technologies and a move to information and service-based activities combine with a growing ability and willingness to improve environmental protection (Lindmark, 2002; Munasinghe, 1999).

While the EKC curve is a good model for the link between environmental change and income growth, this is not necessarily applicable to all types of air pollutants (Yandle et al, 2004). Ozone (O_3), nitrogen oxides (NO_x) and particulate matter (PM) are pollutants of concern (Lenzen et al, 2006). In addition, improvements in air quality with income growth are not inevitable but are dependent on government policies, social institutions and the functioning of the economy. Therefore, it is important to establish appropriate institutions as well as transboundary approaches to air pollution (Arrow et al, 1995). In countries where the general public have political power, and civil as well as economic rights, air quality improves (Torras and Boyce, 1998).

Development of urban air pollution

Many Asian governments have recognized air pollution as a key environmental problem that needs to be addressed. Those cities that have been able to introduce emission control early in their development path (e.g. Hong Kong, Tokyo and Singapore) have avoided the extremely high levels of urban pollution that are often associated with other cities that have introduced emission control measures later. The earlier integrated AQM systems are introduced, the lower the maximum pollution levels that will occur (UNEP/WHO, 1992). Figure 1.2 relates the development of air pollution problems in cities to their development status (UNEP/WHO, 1992). This general developmental path is one that is current in many cities around the world.

While conventional wisdom expects air pollution to increase as Asian countries undergo economic development, this is not the case for all countries and pollutants. Since the 1990s, sulphur dioxide (SO_2) emissions in Asia have declined, mostly due to a reduction of SO_2 emissions in China (Sinton and Fridley, 2000). China has decreased its emissions of SO_2 from 23.8 million tons in 1995 to 20 million tons in 2000. This decrease is due to a general reform of industry and power generation including a substantial decline in industrial high-sulphur coal use and an improvement in energy efficiency and economic growth (Cofala et al, 2004). The reduction of particular pollutants (e.g. NO_x, PM and O_3) has been slow in some Asian countries due to an increase in the number of vehicles, which offsets the emission reductions achieved by improved vehicle technology.

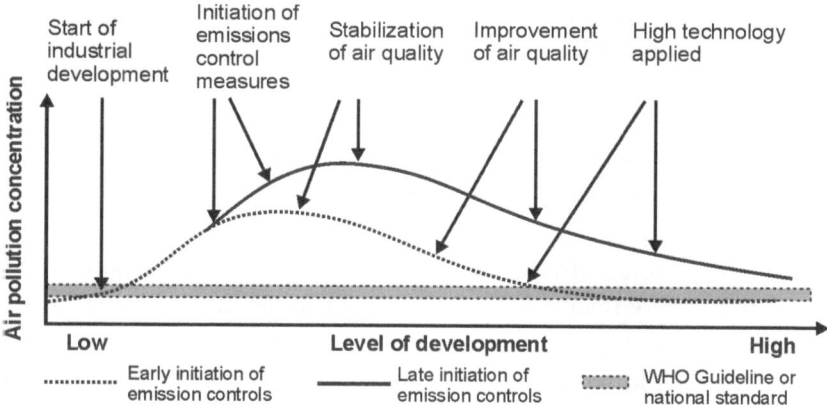

Figure 1.2 *Development of air pollution problems in cities*

Source: based on UNEP/WHO (1992)

UNDERSTANDING URBAN AIR POLLUTION

As the wealth of a city increases so do the key driving forces. These are mainly road transport, power and heat production, industry and agriculture. In some countries such as China, where climatic conditions require space heating at least in the northern provinces, domestic air pollution may be significant in the urban and rural areas. These sectors create a pressure on the environment in the form of emissions of air pollutants, which affect the state of air quality.

Once released into the atmosphere, air pollutants undergo chemical reactions resulting in a wide variation in pollutant concentrations with time and location and corresponding exposure of the population. Pollutants such as SO_2, NO_x and carbon monoxide (CO) typically occur in high concentrations close to their sources (e.g. industrial plants and streets) and show low concentrations elsewhere (EEA, 2003). A secondary pollutant such as O_3 and the deposition of acidic compounds can occur over larger areas (both rural and urban) which can result in high regional background concentrations (Vingarzan, 2004). Particulate matter ($PM_{2.5}$ and PM_{10}) can also show a rather high regional background level (Hien et al, 2004). Particulate matter (PM) in this context refers to the sum of primary PM and the weighted emissions of secondary PM precursors. Primary PM refers to particles emitted directly to the atmosphere. Secondary PM precursors are pollutants that are partly transformed into particles by photochemical reactions in the atmosphere. In some countries, PM background concentrations may become extremely high at certain times, e.g. when sandstorms occur in China and move to urban areas (UNEP, 2005).

Air pollution has an impact on human health, ecosystems and materials with corresponding economic losses. The response of many national and city authorities has been air quality regulations such as the regulation of the sources of pollutant emissions and the monitoring of air quality against specific air quality standards and guidelines. Historically, the combination of driving forces and pressures seldom overwhelmed the natural resilience of urban ecosystems in the early stages of economic development. However, many Asian cities now have to suffer the pressure of a combination of different driving forces, which are occurring simultaneously, each with a greater intensity than has occurred elsewhere or in the past (McGranahan et al, 2001). Therefore many cities have to make greater efforts to cope with the combined pressures. While reductions may be achieved for certain pollutants, others are becoming more difficult to address due to the absence of a well-developed infrastructure, integrated planning and the financial resources to restore environmental quality.

A causal framework to describe these interactions in society is the Driving Forces, Pressures, States, Impacts, Responses (DPSIR) framework, which can be used to understand urban air pollution in Asia (EEA, 1999) (see Figure 1.3).

DRIVERS OF AIR POLLUTION IN ASIA

The Asian region has experienced a rapid growth in population and urbanization over the last decade, especially within South Asia. The percentage of the population living in cities is expected to substantially increase over the coming years. The growth in population, urbanization and a strong economy has been accompanied by a rapid growth in motorization and a considerable increase in energy use. The cumulative impact of these different driving forces has been experienced in some of the more populous countries in Asia such as China and India.

Demography and urbanization

Asian cities have grown very rapidly as Asia's population increased from 2.4 billion in 1975 to 3.8 billion in 2003 (see Figure 1.4) and many cities will continue to grow. Overall, population growth has slowed to the world average of 1.3 per cent a year. However, Asia's share of the global urban population has risen from 9 per cent in 1920 to 48 per cent in 2000 and is expected to rise to 53 per cent by 2030 (UN-HABITAT, 2004).

Economic development in many Asian countries has caused rural-to-urban migration with people from the village moving to the city in the hope of a better life. In the mid-1960s approximately one person in five lived in a city. In the mid-1990s the ratio became one person in three. It is expected that by 2020, approximately one in two

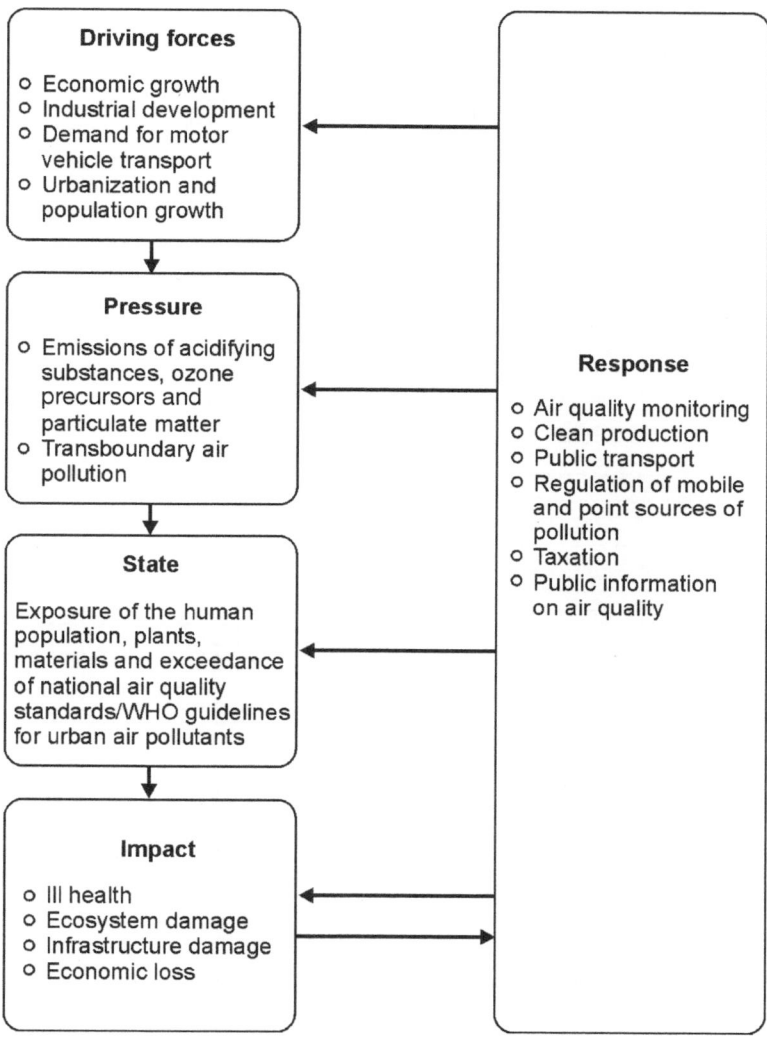

Figure 1.3 *The Driving Force-Pressure-State-Impact-Response (DPSIR) Framework*

Source: EEA (1999)

people in Asia will be living in a city. An estimated 11 megacities with a population of 10 million or greater – as of 2003 – exist in the Asian region (Beijing, Delhi, Dhaka, Jakarta, Karachi, Kolkata, Metro Manila, Mumbai, Osaka-Kobe, Shanghai and Tokyo) with Greater Tokyo being the largest urban conurbation with a population of approximately 35 million (UN, 2004).

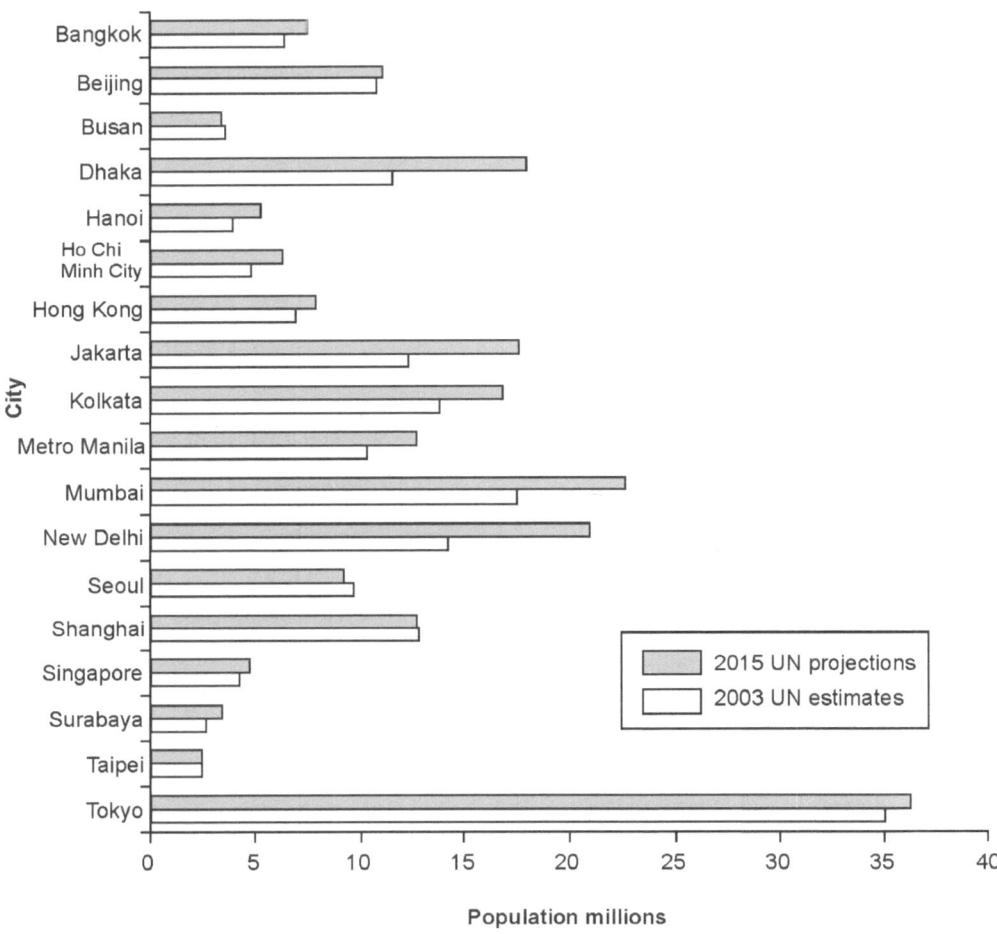

Figure 1.4 *Estimated population of selected Asian cities in 2003 and projected population in 2015*

Source: UN (2004)

Economic development

Developing countries in Asia have experienced economic development and higher standards of living over the past 30 years; this is especially true for China, India and South Korea. In 2002 real gross domestic product (GDP) in Asia grew by 5.6 per cent compared with 3.9 per cent in 2001. The economic structure of the region has changed, with the importance of agriculture diminishing and the service sector growing.

In the period 2001–2025 the highest growth in world economic activity is expected to occur in Asia. Regional GDP in Asia is predicted to increase by 5.1 per cent per year compared with 3 per cent per year for the world as a whole. China is expected achieve the highest annual growth of 6 per cent per year placing it, by 2023, among the world's major economies behind Japan and the US (IEA, 2004).

Energy consumption

The growth in GDP has resulted in increasing levels of energy consumption. Primary energy consumption in the Asia Pacific region has been steadily increasing over the past 30 years, reflecting the economic development of the region (BP, 2004).

Asia is responsible for 23 per cent of world energy consumption with China having a high share at 13 per cent. During the period 2002–2003, China accounted for 57 per cent of the increase in world energy consumption. The growth in Chinese energy demand was approximately 11 per cent in the period 2002–2003 (Enerdata, 2004). Coal is a dominant source of energy in Asia, with both China and India estimated to be responsible for approximately 67 per cent of the increase in coal use worldwide (IEA, 2004). Constrained refining capacity combined with a large demand for gasoline and diesel is expected to result in an increase in coal use, especially in the power sector. This might offset the recent reductions achieved in SO_2 emissions in China if no desulphurization measures are taken.

Population growth and urbanization in China and India is expected to result in a continued increase in energy use over the coming decades. Also, increases in residential energy use, rising incomes and rural electrification are expected to result in a high demand for electrical appliances (e.g. air conditioning, refrigeration and laundry equipment) which currently are not widely used (IEA, 2004).

Asian economies have become more efficient in their consumption of supplied energy (less energy intensive) over the period 1990–2000. The energy use per unit of GDP, which indicates the extent to which economies are efficient in their consumption of supplied energy, has decreased, reflecting the global trend. Nevertheless, the increasing overall consumption of coal, natural gas and oil has resulted in higher levels of emissions of certain pollutants. China aims to quadruple GDP from 2000 to 2020 while only doubling energy use. This is seen as a challenge to energy security, social welfare and environmental objectives. At the same time, the country seeks to move towards cleaner fuels in all sectors. However, more recently, energy use has been rising faster than GDP indicating energy inefficiency. The dependence on coal shows no sign of diminishing and imports of oil and gas are set to rise in the future (Sinton et al, 2005).

Motorization

The level of motorization in Asian countries has been growing at different rates and has been influenced by increasing affluence and population growth. The number of vehicles in Asia will continue to grow, with the possible exception of Singapore and Hong Kong, where active controls have been implemented to limit the number of vehicles and promote alternative public transport options.

The majority of additional vehicles will be new ones, while for some countries a substantial number of reconditioned vehicles, for which there is no adequate emission control in place, will be added to the fleet. These are often buses or other public transport utility vehicles imported from other Asian countries where stringent in-use emissions and safety requirements have made the engines obsolete. These are typically vehicles which are used intensively and not maintained very well. At present, vehicles in major industrialized Asian countries are being used for longer than was considered would be their normal useful life.

Two-wheelers play an important role in Asia as they are accessible to a larger portion of the population who cannot afford to own a motor vehicle. This region has a very large population of two- and three-wheelers that dominate the emissions inventory for some compounds in most major cities. More than 75 per cent of the world's fleet of two-wheelers and three-wheelers is in Asia. China alone accounts for approximately 50 per cent and India for 20 per cent of the world's fleet of two-wheelers (WBSCD, 2004). In some cases, the inclusion of motorized two-wheelers in total motorization rates brings these motorization rates up to the same level as cities with much higher average per capita (or per household) incomes. For example, when motorized two-wheelers are included, Mexico City, with a GDP per capita 10 times higher than Chennai, has a lower motorization rate than Chennai. As the income of individuals rises, the two-wheeler becomes accessible and therefore accelerates the motorization process in Asia.

A high proportion of the population in Asian cities currently depends on public transport and non-motorized forms of transport such as walking and cycling as the main mode of transport. As cities grow in area and decline in average density, the relative share of trips made by public transport and the absolute number of non-motorized trips are decreasing.

The rising trend in car ownership and decline in public transport as an alternative to the private motor vehicle has varying effects on different countries often affecting the poor, elderly and disabled who are dependent on public transport for their personal mobility (WBSCD, 2004). The continued growth in motorization in Asia will further exacerbate current levels of traffic congestion and result in increased motor vehicle emissions.

PRESSURE OF AIR POLLUTION IN ASIA

Pollutant emissions

Economic development, together with population growth, urbanization, motorization and high levels of energy consumption, has resulted in pressures on the environment in the form of pollutant emissions. Local sources of air pollution include:

- small sources such as boilers, generators, light industries
- point sources such as major industrial sites
- mobile sources such as on-road motor vehicles
- area sources such as waste deposits and places where open burning of waste materials from agriculture, forestry and land clearance occur
- local biogenic or natural sources such as deserts and eroded areas
- transboundary air pollution from distant sources comprising emissions from fixed, mobile and other sources.

Total anthropogenic emissions in Asia for the year 2000 show that the power, industry and transport sectors are major contributors to key urban air pollutants (Streets et al, 2003a). In particular, China is a significant contributor to emissions of PM, SO_2, NO_x, CO and CO_2 in the region, reflecting the rapid economic development that has been taking place in this country (see Figure 1.5). From studies in many Chinese and Indian cities it is evident that these cities are major sources of PM (Wang et al, 1999; 2005; Streets et al, 2001; Bi et al, 2002; Duggal and Pandey, 2002; Cao et al, 2003; Fu, 2004; Li et al, 2004; NEERI, 2004; Sun et al, 2004; Zheng et al, 2004; Bhattacharya, 2005). PM is also emitted from desert areas during periods of higher wind speed, vegetation fires and agricultural burning, and through chemical processes during transport of gases. South Asia has the second highest PM emissions in the region. O_3 is also produced through the transformation of NO_x and hydrocarbons; however, O_3 levels are mostly unknown as monitoring of O_3 is limited in Asia.

In 2000 China was estimated to have produced approximately 45 per cent of total Asian SO_2 emissions due to the use of coal-fired power stations and 36 per cent due to the industrial sector. The transport sector is estimated to have produced 37 per cent of NO_x while power generation produced 27 per cent in Asia. Estimates for the domestic sector suggest that it is responsible for 64 per cent of black carbon (BC) and 65 per cent of organic carbon (OC) emissions from the combustion of coal, kerosene and biofuels. The residential sector (about 34 per cent) and the transport sector (about 27 per cent) were key contributors to emissions of non-methane volatile organic compounds (NMVOC) in Asia in 2000 (Street et al, 2003b).

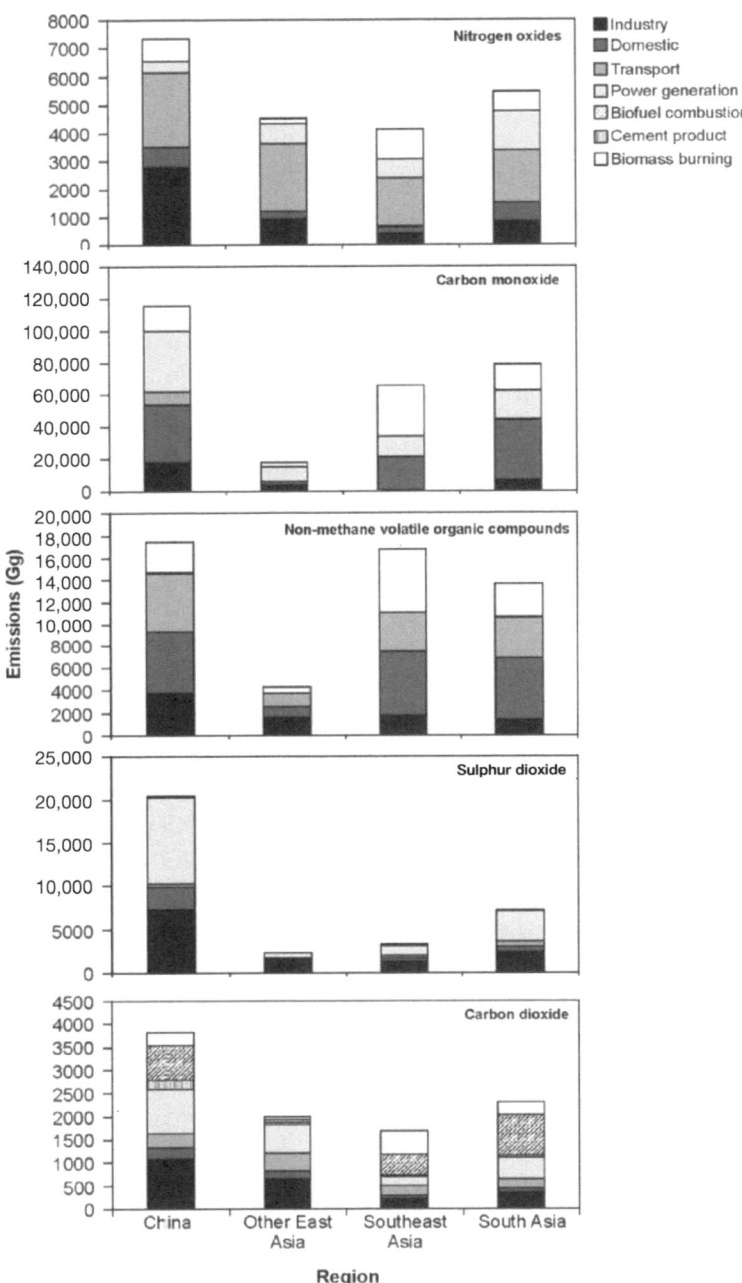

Figure 1.5 *Total anthropogenic emissions in Asia for selected pollutants for the year 2000*

Source: Streets et al (2003a)

A study of the sources of fine particulate pollutants in New Delhi, Kolkata and Mumbai indicated that there is no single dominant source. Sources of PM differ by location and season among the three Indian cities examined (World Bank/ESMAP, 2004). For example, mobile sources and biomass combustion contribute substantially to PM emissions and in several cases approximately in equal proportions (spring and autumn in Delhi and autumn in Mumbai). The contribution of road dust can also be significant (summer in Delhi and spring and autumn in Mumbai). The combined contribution of biomass and coal is highest in winter in New Delhi and Kolkata as a result of heating. Contributions from solid fuel combustion are also significant in non-heating seasons: in spring and autumn in New Delhi and in autumn in Mumbai, probably due to the use of solid fuels in small-scale industries and by households for cooking. The study concluded that vigorously pursuing control measures in one sector alone (e.g. transport), while leaving other sectors largely untouched, is less likely to result in a marked improvement in urban air quality than if a multi-pronged approach addressing a number of sources and sectors is adopted (World Bank/ESMAP, 2004).

Greenhouse gases

Fossil fuel, such as coal, is a predominant fuel in Asia's developing countries. It is expected to continue to be so as levels of economic growth and energy consumption increase. By 2020 CO_2 emissions from developing countries, especially China and India are expected to surpass those in the industrialized countries. CO_2 emissions in developing countries in Asia are expected to increase from 6012 million metric tons in 2001 to 11,801 million metric tons in 2025 (IEA, 2004).

The Kyoto Protocol sets targets for industrialized countries to cut their greenhouse gas emissions; it became legally binding on 16 February 2005. It requires a 5.2 per cent cut in greenhouse gas (GHG) emissions from the industrialized world as a whole by 2012. In 2001, China and India together accounted for 17 per cent of total world CO_2 emissions compared with the US at 24 per cent. Although both countries account for a significant amount of the world's CO_2 emissions, they have no commitment under the Kyoto Protocol as they are classed as developing nations (IEA, 2004). However, the most rapid improvement in CO_2 intensity is expected to occur in developing countries in Asia. As the economy develops, there will be an improved efficiency of capital equipment and an increased use of less carbon-intensive fuels such as natural gas for new electricity generation. In particular, China is expected to remain heavily dependent on fossil fuels, yet CO_2 intensity is expected to decrease by 2.6 per cent during the period 2001 to 2025 (IEA, 2004; IGES, 2004).

Transboundary air pollution

Urban air quality in Asia is becoming increasingly affected by pollution resulting from neighbouring jurisdictions. Transboundary air pollution in the form of atmospheric brown cloud, acidification and dust and sandstorms all contribute directly or indirectly to effects on human health, environment and the economy in the region (Lam et al, 2005; Min, 2001).

The high levels of direct and indirect GHG emissions such as SO_2, NO_x, CO, NMVOC and ammonia (NH_3) are strongly linked to the consumption of fossil fuels and have also contributed to the problem of transboundary air pollution within the region.

Atmospheric brown cloud

The atmospheric brown cloud (ABC) is a regional phenomenon of aerosol and PM plumes which occurs across parts of Asia. In 2001 the Indian Ocean Experiment (INDOEX) documented the vast extent of a 3 kilometre thick brownish layer of pollutants hovering over most of tropical Indian Ocean, South, Southeast and East Asia. The haze consists of sulphates, nitrates, organics, black carbon and fly ash among several other pollutants, which can be transported far beyond their source region, particularly during the dry season. The potential direct effects include a significant reduction in the solar radiation reaching the surface; a 50–100 per cent increase in solar heating of the lower atmosphere; rainfall suppression; agricultural productivity decline; and adverse human health effects (UNEP, 2002).

The indirect effects include cooling of the land surface, increase in frequency and strength of thermal inversions which trap more pollution; evaporation reduction; soil drying; and disruption of monsoon rainfall. The regional and global impacts of the ABC are set to intensify over the next 30 years as the population of the Asian region and sources increase.

Southeast Asia haze

Forestland conversion involving uncontrolled use of fire in land preparation has been identified as a major source of fires in the Southeast Asian region. Transboundary air pollution caused by these fires, used to clear land in Indonesia, remains a perennial problem. In the years when the El Niño phenomenon returns to the Southeast Asian region with its drier conditions, the worsening haze from forest and brush fires becomes an increasing challenge for adjacent countries. The impacts of the haze include major disruption of travel and transport, increasing risk of traffic accidents, public health risks through the transboundary transport of fine and ultrafine PM and environmental damage. PM pollution levels during the episode in 1997/1998 were extremely high in Indonesia (more than 1 mg/m^3 in Sumatera) and high (several hundreds µg/m^3) in Malaysia (Frankenberg et al, 2005; Sastry, 2002).

As a consequence of the 1997/1998 fire event, the Association of Southeast Asian Nations (ASEAN) and the Asian Development Bank (ADB) developed a Regional Haze Action Plan (RHAP). The primary objectives of the RHAP are to: (i) prevent forest fires through better management policies and enforcement; (ii) establish operational mechanisms to monitor land and forest fires; and (iii) strengthen regional land and forest firefighting capacity with other mitigation measures (ADB/ASEAN, 2001). At the same time WHO, in collaboration with UNEP and the World Meteorological Organization (WMO), developed the Health Guidelines for Vegetation Fire Events (WHO/UNEP/WMO, 1999). These guidelines emphasized the global occurrence of fires, analysed their causes, discussed the methods of observation and developed an early warning system for vegetation fire events with respect to the protection of public health.

In 2002, ASEAN adopted the ASEAN Agreement on Transboundary Haze Pollution to prevent and monitor transboundary haze pollution as a result of land and/or forest fires, and to control sources of fires (ASEAN, 2002).

Regional acidification

The rapid growth of cities in Asia has resulted in greatly increased emissions of SO_2 and NO_x. Acid deposition can take the form of wet precipitation, such as rain, fog, mist or snow, or dry deposition involving the deposition of acidic gases and aerosols. Acid deposition, also called 'acid rain', is caused mainly by SO_2 and NO_x, NH_3 emissions from fossil fuel combustion, industrial processes and agricultural practices. Acid wet deposition refers to rainwater with a pH of 5.6 or less. In the 1990s, acid rain as severe as that in Europe and North America was widely observed in Japan, with pH values as low as pH 4.3 (MoE Japan, 1998; EEA, 2005). Acid deposition has far-reaching impacts. It affects fish and other aquatic biota due to the progressive acidification of inland waters, including marshes, lakes, rivers and streams. Acid deposition threatens forests due to soil acidification. It accelerates and aggravates the decline and decay of monuments and buildings of cultural importance.

SO_2, NO_x and NH_3 have only short residence times in the atmosphere (1–5 days) and can, therefore, be transported only over a few hundred kilometres before becoming deposited or transformed into secondary fine particles. Effects of these gaseous compounds include impacts on human respiratory health; adverse impacts on forest and crop growth, particularly in sensitive soils; corrosion of human-made structures; and adverse changes in natural ecosystems. The deposition of sulphur in some areas of Asia is now equal to or higher than in the most polluted areas of Europe and

North America. Soil acidification and acidification-induced vegetation damages have been observed in parts of China (Seip et al, 1999). In addition, acid aerosols such as sulphates, nitrates and ammonium aerosols have a much larger residence time allowing for transport of these particles over several thousand kilometres. In order to cope with increasing risks of problems related to excess deposition of acidic substances, the Acid Deposition Monitoring Network in East Asia (EANET) was established as a regional cooperative initiative to promote efforts for environmental sustainability and protection of human health. The ten countries participating in the preparatory phase activities, namely, China, Indonesia, Japan, Malaysia, Mongolia, Philippines, Republic of Korea, Russia, Thailand and Vietnam, decided to start the EANET activities on a regular basis from January 2001. Cambodia and Laos joined the network in November 2001 and November 2002 respectively, with monitoring activities being started in these countries in 2003 (EANET, 2005a). Monitoring activities include wet deposition, dry deposition and assessment of adverse impacts of acidic substances on the vegetation, the soil and the inland aquatic environment. Table 1.1 gives an overview of variables considered in the project. Since 2001 data reports on the Acid Deposition in the East Asian Region have been published and made available on the internet (EANET, 2005b).

Hemispheric transport of air pollutants

While local and regional emissions sources are the main cause of air pollution problems worldwide, there is increasing evidence that air pollutants are transported on a hemispheric or global scale. These include:

Table 1.1 *Monitored variables of EANET*

Wet deposition	pH, EC, concentrations of ions (SO_4^{2-}, NO_3^-, Cl^-, NH_4^+, Na^+, K^+, Ca^{2+}, Mg^{2+}, etc.)
Dry deposition	Gas concentrations (SO_2, NO/NO_2, O_3, HNO_3, NH_3, etc.)
Soil	pH, CEC, concentrations of exchangeable ions, etc.
Vegetation	Degree of decline of trees, abnormalities of leaves and branches, etc.
Inland aquatic environment	pH, EC, alkalinity etc.

Notes: EC: electric conductivity: CEC: cation exchange capacity.

Source: EANET (2005b)

- ozone and its precursors
- fine and ultrafine particles
- acidifying substances
- mercury
- persistent organic pollutants.

In the northern hemisphere, the hemispheric transport of air pollutants (HTAP) may be important for understanding air pollution problems in urban areas. In 2005, the Executive Body of the United Nations Economic Commission for Europe (UNECE) Convention on Long-range Transboundary Air Pollution (LRTAP) established a Task Force on Hemispheric Transport of Air Pollution to examine this issue further (UNECE, 2004; LRTAP, 2005).

Dust and sandstorms

The Asian region has a wide range of terrain and climates. Much of it is dry land but some of it is desert. Many countries within the Asian region are experiencing severe and, in some cases, accelerating desertification. High population density, low levels of economic development, poor infrastructure and isolation exacerbate the problems. Sand and duststorm events are frequent and serious causing loss of crops, destruction of houses, disruption in transport and communication, as well as putting human health at risk.

Dust from these regions is often distributed beyond the national borders and this poses a problem for neighbouring countries, often directly affecting any improvements in air quality. Dust and sandstorms are now occurring in Northeast Asia nearly five times as often as they were in the 1950s. The storms, which originate in the dry regions of northern China and Mongolia and blow across the Korean peninsula and Japan, are also growing in intensity. In April 2002 dust levels in Seoul, 1200 kilometres away from their source, reached 2070 μg/m^3, almost 14 times the 24-hour PM$_{10}$ level of 150 μg/m^3, the United States Environmental Protection Agency (USEPA) standard. The storms cause considerable hardship though respiratory problems and related deaths, loss of income, disruption of communication and loss of livestock and crops over large areas (UNEP, 2004; 2005).

STATE OF AIR POLLUTION IN ASIA

Some of the highest levels of outdoor air pollution in the world are found in Asian cities. However, limited AQM systems, capacity and resources have meant that monitoring of air quality in some Asian cities has been ad hoc and data on air quality trends and exceedance of national WHO guidelines are not always available.

Using the WHO Air Management Information System (AMIS) it is possible to calculate the average concentrations for selected pollutants over the period 1990–1999. In addition, annual air quality data provided by the 20 Asian cities were averaged to provide data for the period 2000–2004. Air pollutant concentrations for 2000–2004 compared with AMIS data (WHO, 2001) shows that there has been a general improvement in air quality (see Figure 1.6). In 4 of the 13 cities, which are covered by AMIS and this study, air quality conditions have worsened for SO_2, NO_2 and PM_{10}. All the cities recorded an improvement in total suspended particulate (TSP) levels. For some cities, the data are insufficient to allow comparison (see Table 1.2). The data presented in Figures 1.6, and in Table 1.2 have to be interpreted with caution since monitoring stations even of the same type (industrial, commercial, residential, kerbside) may have different definitions in different cities and monitoring may not be of the same quality.

Table 1.2 *Air quality data for 2000–2004 compared with 1990–1999*

	SO_2	NO_2	SPM	PM_{10}
Bangkok	≥	>	<	<
Beijing	<	?	?	<
Busan	<	>	?	≥
Colombo	>	<	?	≤
Hong Kong	>	<	<	<
Kolkata	<	>	<	<
Manila	?	?	<	?
Mumbai	<	<	<	<
New Delhi	<	<	<	<
Seoul	<	>	<	<
Shanghai	<	<	<	?
Taipei, China	<	<	?	>
Tokyo	<	≤	?	<

Notes:
> (<) = Data for 2000–2004 larger (smaller) by more than 5 per cent than data for 1990–1999.
≥ = approximately 5 per cent increase.
≤ = approximately 5 per cent decrease.
? = no data available.

Source: WHO (2001), CAI-Asia (2006c)

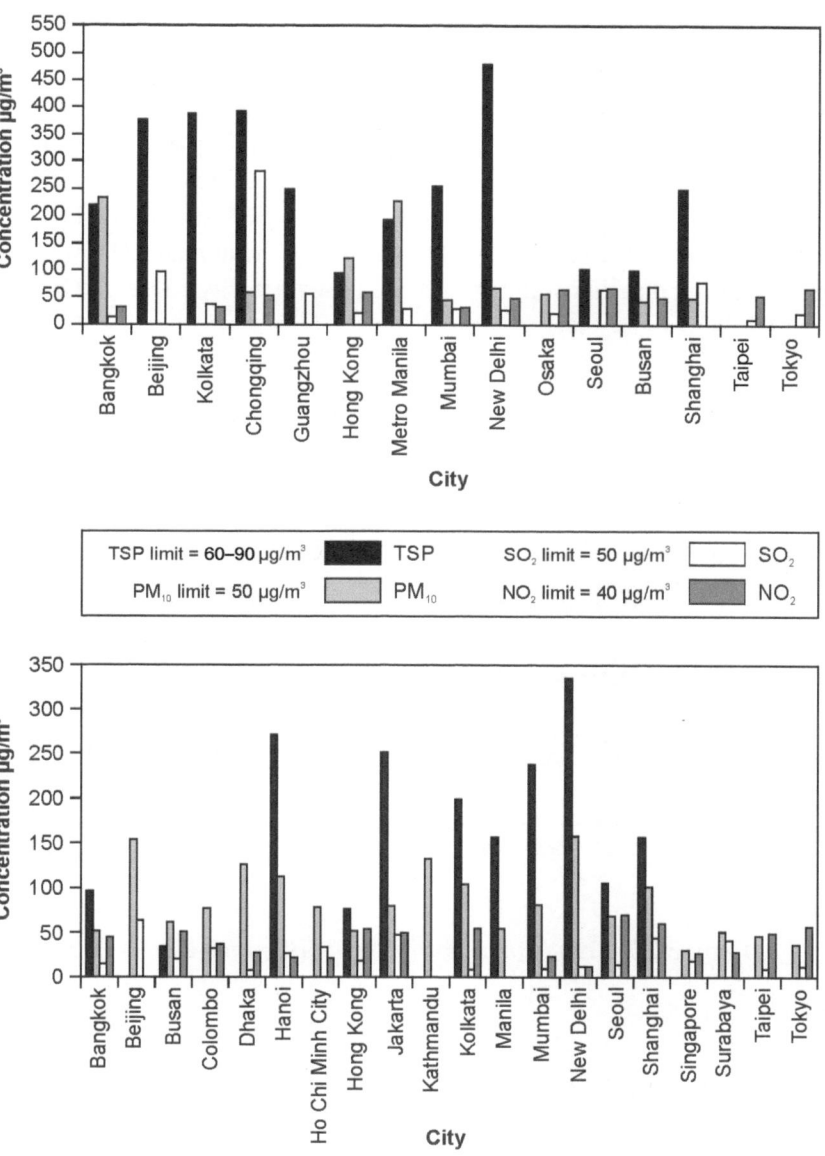

Figure 1.6 *Average air quality levels of selected Asian cities, 1990–1999 (AMIS) and 2000–2004*

Note: (top) 1990–1999, (bottom) 2000–2004.

Source: CAI-Asia (2006c)

Figure 1.7 presents a macro-analysis of the air pollution trends in the 20 cities. Air quality data for O_3 and NO_x levels are insufficient to formulate clear trends for the period 1994–2004. The number of cities monitoring these pollutants is still relatively limited, especially in the case of O_3, while the quality of monitoring is varied, especially for NO_x. In general, TSP and PM_{10} have decreased in the period 1994 to 2004. However, ambient levels remain above USEPA standards. In contrast, the aggregated data for SO_2 clearly show that it is below the annual EU limit and USEPA standard indicating that this pollutant has been satisfactorily addressed by most Asian cities.

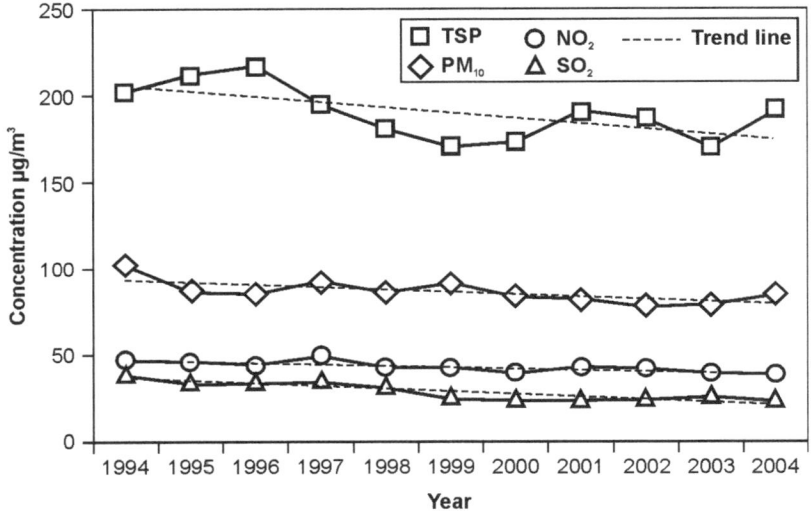

Figure 1.7 *Trends in major criteria pollutants (1994–2004)*

Source: CAI-Asia (2006c)

IMPACTS OF AIR POLLUTION IN ASIA

The impact of air pollution has a range of direct and indirect effects on human health, ecosystems, vegetation and material assets. The spatial scale of pollution impacts ranges from the street level to urban, peri-urban and rural areas. The timescale of effects also varies ranging from hours to years. Table 1.3 presents the issues associated with the impact of air pollution.

Human health

In a comparative risk assessment of global human health risks, WHO ranked urban outdoor air pollution as the tenth leading cause of premature deaths and indoor air

Table 1.3 *Impacts of air pollution*

Issue	Spatial scale	Effects-related timescale
Human health-related impacts due to exposure to ozone and particles (and to a lesser extent to NO_2, SO_2, lead and benzene)	Urban areas, streets Ozone: also rural areas	Hours, days, years
Acidification and eutrophication of water, soils and ecosystems	Range: 100–1000 km	Years
Damage to vegetation and crops due to exposure to ground-level ozone	Range: 100–1000 km	Hours, growing season
Damage to materials and cultural heritage due to exposure to acidifying compounds and ozone	Urban, rural areas	Years

Source: EEA (2003)

pollution as the fourth leading cause (WHO, 2002). WHO estimate that 537,000 persons die prematurely each year in the Southeast Asia and Western Pacific regions due to exposure to urban air pollution. Even more people, approximately 1.6 million, die annually from exposure to indoor air pollution in these regions. In addition, millions of people are affected by respiratory diseases, which are either caused or aggravated by exposure to ambient air pollution (WHO, 2002).

High levels of outdoor air pollution from stationary and mobile sources contribute considerably to the burden of disease from disorders of the cardiovascular and respiratory systems. In addition, indoor air pollution resulting from domestic use of solid and fossil fuels, particularly in rural areas, account for a large number of diseases (including respiratory infections and chronic obstructive pulmonary disease), especially among young children and women (WHO, 2002).

WHO estimate that two-thirds of the world's statistics in deaths and lost life years due to urban outdoor air pollution occur in Asia (WHO, 2002). These estimates, however, are based largely on exposure–response relationships derived in Europe and North America before 2000 and extrapolated for Asia (WHO, 2000a; 2000b). Since the publication of the air quality guidelines for Europe (WHO, 2000b) a large body of evidence has emerged on the effects PM and O_3 on human health. In 2003 WHO reviewed the health effects of PM, O_3 and NO_2 in Europe. A meta-analysis of European timescale studies of PM and O_3 was undertaken, which resulted in the

Table 1.4 *Percentage change estimates for a 10 μg/m³ increase in PM$_{10}$, black smoke and O$_3$ and all-cause, respiratory and cardio-vascular mortality, all ages (95% confidence intervals)*

Outcome disease	Number of estimates	PM$_{10}$	Number of estimates	Black smoke	Number of estimates	O$_3$ (8-hour)
All-cause mortality	33	0.6 (0.4, 0.8)	33	0.4 (0.2, 0.7)	20	0.20 (0.05, 0.35)
Respiratory mortality	20	1.0	24	–0.1 (–1.0, 0.8)	15	–0/1 (–0.5, 0.4)
Cardio-vascular mortality	23	0.5 (0.1, 1.0)	22	0.4 (0.1, 0.6)	17	0.4 (0.3, 0.5)

Source: WHO (2004)

revision of the relative risk estimates for a 10 μg/m³ increase in PM$_{10}$, black smoke and O$_3$. Table 1.4 presents the revised estimates, expressed in terms of percentage changes in health outcomes for a 10 μg/m³ increase in PM$_{10}$, black smoke and O$_3$.

Compared to Asia, Europe and North America differ in the type and exposure to air pollution and the health status of the population as well as the provision and accessibility to health care systems. These differences give rise to uncertainties when the results of health impact studies from more developed countries are applied to developing countries in Asia. In 2002 the Clean Air Initiative for Asian Cites and the Health Effects Institute initiated the Public Health and Air Pollution in Asia (PAPA) programme to reduce these uncertainties.

The PAPA programme conducted a comprehensive literature review of studies on health impacts of air pollution in Asia. The review covered 138 studies that were published in peer-reviewed literature from 1980 to 2003. These studies were also conducted mostly in East Asia (mainland China, Taipei, Hong Kong, South Korea and Japan) with a few conducted in South and Southeast Asia (HEI, 2004).

Results of the literature review showed that summary estimates of PM$_{10}$ and SO$_2$, which is approximately a 0.4 to 0.5 per cent increase in all-cause mortality for every 10 μg/m³ of exposure, resemble the values reported by the large-scale European and North American studies that used comparable methods (see Table 1.5).

The interaction between air pollution, health and poverty in Asia is important, especially in areas with high air pollution and large populations at the lowest socioeconomic levels. The potential public health and social policy implications are significant, yet they are not well understood. While there have been several studies which document the overall health impact of air pollution in Asian cities, there have been very few studies on the relationship of air pollution, health and poverty

Table 1.5 *Summary of estimates of percentage change[a] in health outcomes[b]*

Outcome	Pollutant	Number of estimates	Heterogeneity[c]	Fixed-effects estimates	Random-effect estimates	Test of publication Bias[d]	Multiplicity study summary estimates (95% CI)
All-cause mortality	PM_{10}	4	0.14 (0.25, 0.56)	0.41	0.49 (0.23, 0.76)		APHEA 2[e] 0.6 (0.4, 0.8) NMMAPS[f] 0.41 (0.29, 0.53)
	TSP	10	0.55	0.20 (0.14, 0.26)	0.20 (0.14, 0.26)	0.53	
	SO_2	11	<0.001	0.35 (0.26, 0.45)	0.52 (0.30, 0.74)	0.03	APHEA 1[g] 0.40 (0.3, 0.5)
Respiratory admissions	NO_2	4	<0.01	0.28 (0.09, 0.47)	0.95 (−0.05, 1.94)		
	SO_2	4	0.03	0.07 (−0.28, 0.41)	0.16 (−0.46, 0.77)		

Notes: [a] per 10 μg/m^3 increase in ambient pollutant concentration.

[b] calculated when four or more studies provided estimates for individual pollutant–outcome pairs.

[c] P value for χ^2 test (P values < 0.05 were considered statistically significant).

[d] P value from Begg test. The test was not conducted for those pollutant–outcome pairs with too few estimates (P value < 0.05 were considered statistically significant).

[e] 29 European cities (Katsouyanni et al, 2001).

[f] 90 US cities (Samet et al, 2000).

[g] 12 European cities (Katsouyanni et al, 1997).

Source: HEI (2004)

in developing countries in general, and in Asia in particular (HEI, 2004). There is emerging evidence that the poor are affected disproportionately by ambient air pollution due to greater exposures, weaker biological defence mechanisms against air pollution and more limited abilities to seek medical assistance once affected by air pollution (Martins et al, 2004; Gouveia and Fletcher, 2000; Stern, 2003).

The poor are subject to higher exposures of air pollution and may be more susceptible to air pollution due to inadequate nutrition, access to medical care, and other factors. At the same time, air pollution could exacerbate the conditions of poverty. It is also expected that the poor will not seek medical assistance quickly, nor be able to afford medicines and treatment. In Hong Kong, it is estimated that the poor suffer increased exposure to air pollution and those in low-income areas are five times as likely to be hospitalized as those residing in high income areas (Stern, 2003). The economic consequences of an increased burden of disease due to air pollution, including cost of illness and loss of income, could also be substantial.

High levels of outdoor air pollution from stationary and mobile sources contribute considerably to the burden of disease with disorders of the cardiovascular and respiratory systems. In addition, indoor air pollution resulting from domestic use of solid and fossil fuels, particularly in rural areas, accounts for a large number of cases of several diseases (including respiratory infections and chronic obstructive pulmonary disease), especially among young children and women (WHO, 2002). A number of key air pollutants have the most deleterious effect on human health. These include PM, O_3, CO, SO_2 and NO_2. A brief description of the health effects of each of these pollutants is given below (Schwela, 2000; WHO, 2000b; 2003; HEI, 2004).

Particulate matter

Health effects of PM in humans depend on particle size and concentration, and can change with daily fluctuations in PM_{10} or $PM_{2.5}$ levels. They include acute effects such as increased daily mortality, increased rates of hospital admissions for exacerbation of respiratory disease, increases in the prevalence of bronchodilator use and coughing, and peak expiratory flow reductions. An extensive body of experimental and epidemiological literature demonstrates that significant associations between short-term and long-term exposure to concentrations of PM and rates of mortality and morbidity exist in the human population. Relationships between PM_{10} or $PM_{2.5}$ exposure and acute health effects appear to be linear at concentrations below 100 µg/m^3. Current epidemiological time-series studies do not indicate a threshold below which no effects occur. Rather they suggest that even at low levels of PM, short-term exposure is associated with daily mortality, daily hospital admissions, exacerbation of respiratory symptoms, bronchodilator use, cough and peak expiratory flow. WHO performed a meta-analysis of existing European time-series studies and estimated relative risks and their 95 per cent confidence intervals for a 10 µg/m^3 increase in

PM_{10}, black smoke and O_3 for all-cause and cause-specific mortality (see Table 1.4) and for respiratory hospital admissions (see Table 1.6).

A small number of studies refer to the long-term effects of PM with respect to mortality and respiratory morbidity. However, the database on health impacts due to long-term exposure to PM has been expanded by new cohort studies and the extension and re-analysis of existing ones (WHO, 2003; HEI, 2003). Strong evidence on the effect of long-term exposure to PM on cardiovascular and cardiopulmonary mortality can be inferred from these studies. For all-cause and cardiopulmonary deaths, statistically significant increased relative risks were found for $PM_{2.5}$. Cardio-pulmonary and cardiovascular mortality was strongly associated with sulphates and $PM_{2.5}$. Life expectancy may be shortened due to long-term exposure to traffic-related air pollution (WHO, 2003). The associations of morbidity and lung-function parameters with long-term exposure to particles and gaseous compounds were studied in various North American and European cohort and cross-sectional investigations. Significant reductions of lung function parameters (forced expiry volume in 1 s (FEV_1) and forced vital capacity (FVC)) were observed for elevated concentrations of PM_{10}, NO_2 and SO_2. Bronchitis, chronic cough, wheeze and conjunctivitis were also associated with a mixture of pollutants. The strong association between PM_{10}, SO_2, NO_2, and O_3 did not allow the causal nature of the associations to be determined (WHO, 2003; WHO 2000b; Schwela, 2000).

Ozone

Epidemiological studies have shown the effect of short-term exposure to O_3 on mortality and respiratory morbidity. Short-term acute effects of O_3 include decreased

Table 1.6 *Percentage change in estimates of respiratory hospital admissions for a 10 µg/m³ increase in PM_{10}, black smoke and O_3 (95 per cent confidence interval in brackets)*

Age group (years)	No. of estimates	PM_{10}	No. of estimates	Black smoke	No. of estimates	O_3 (8-hour)
15–64	–	–	5	0.6 (0.1, 1.0)	5	0.1 (–0.9, 1.2)
65+	10	0.7 (0.2, 1.3)	7	0.1 (–0.7, 0.9)	5	0.5 (–0.2, 1.2)

Source: WHO (2004)

pulmonary function, increased airway responsiveness and airway inflammation. In addition, pre-existing respiratory diseases such as asthma are aggravated and there is an increase in daily hospital admissions, emergency department visits for respiratory causes, and excess mortality. Exposure–response relationships appear to be non-linear for the associations between O_3 concentration and the lung function parameter (FEV_1), inflammatory changes and changes in hospital admissions, respectively. A linear relationship has been established for the association between the percentage change in symptom exacerbation among adults and asthmatics. In studies on the relationships between O_3 exposure and daily mortality and hospital admissions for respiratory diseases, single-pollutant associations between O_3 and these health effects remained statistically significant even in multi-pollutant models (Schwela, 2000; WHO, 2003, 2000b).

Few studies on the health effects of long-term exposure to O_3 exist. Available evidence suggests that long-term O_3 exposure reduces lung function development in children. A difference of 20 $\mu g/m^3$ in average O_3 exposure was associated with a significant reduction in lung function of 2 per cent. The incidence of asthma in male adults and among children was found to increase with long-term exposure to O_3. The evidence linking exposure to O_3 and asthma prevalence is not consistent. Incidence of mortality due to lung cancer among non-smoking males is found to be strongly associated with long-term concentrations of PM and O_3. Due to a high correlation between PM and O_3, the individual contribution of each pollutant could not be separated. Another cohort study found no such association (WHO, 2003).

Carbon monoxide

CO absorbed in the lungs binds with haemoglobin to form carboxyhaemoglobin, which reduces the oxygen-carrying capacity of the blood and impairs the release of oxygen from haemoglobin. Potential health effects of CO include hypoxia, neurological deficits, neuro-behavioural changes, and increases in daily mortality and hospital admissions for cardiovascular diseases. Several studies have shown small, statistically significant relationships between CO and daily mortality. Other pollutants and other environmental variables, however, were shown to also be significant. There is still a question of whether the demonstrated associations are causal or whether CO might be acting as a proxy for PM. Some studies appear to show that the association between ambient CO and mortality and hospital admissions due to cardiovascular diseases persists even at very low CO levels indicating no threshold for the onset of these effects. It is possible that ambient CO may have more serious health consequences than carboxyhaemoglobin formation, and at lower levels than that caused by elevated carboxyhaemoglobin levels (WHO, 2000b; Schwela, 2000).

Sulphur dioxide

Several single- and multi-pollutant time-series studies have observed associations between SO_2 and daily mortality and morbidity, while others have not. In particular, single-pollutant correlations sometimes disappeared when other pollutants, especially suspended PM, were included. In cross-sectional studies with asthmatics, a particularly susceptible group, significant, non-threshold relationships were found between SO_2 exposure and reduction of the forced expiratory volume in one second (WHO, 2000b; Schwela, 2000).

Nitrogen dioxide

There are indications of weak associations between short-term NO_2 exposure due to gas cooking and respiratory symptoms and lung function parameters in children. This association is not consistently found in women exposed to NO_2 in the short term. A number of outdoor studies indicate that children with long-term exposure exhibit increased respiratory symptoms, decreases in lung function, and increases in the incidences of chronic cough, bronchitis and conjunctivitis. Other long-term NO_2 studies provide little evidence of such effects in adults. In multi-pollutant time-series studies of the association between NO_2 and mortality and morbidity, no association was found, or a single-pollutant association disappeared, when other pollutants were included in the analysis (Schwela, 2000; WHO, 2000b). However, a clear association between personal NO_2 exposure and lung function (FEV_1 and FVC) was found in communities with presumably homogeneous PM concentrations (WHO, 2003).

Agricultural crops

Adverse impacts of air pollution on agricultural crops and tree species around power stations, steel plants, smelters, cement factories and other industries have been documented in several countries in Asia, i.e. China (Zheng and Shimizu, 2003), Taiwan (Sheu and Liu, 2003) and India (Agrawal, 2003). In Pakistan, effects have been observed after exposure to high concentrations of SO_2, NO_x and O_3 alone and/or in combination, while in China field evidence established adverse impacts on crops due to exposure of SO_2, fluorides and PM alone and/or in combination (Wahid, 2003). The air pollutants emitted from a coal-burning power station in Hunan Province resulted in severe impacts on local agricultural productivity and sometimes caused 100 per cent losses in yield (Bao and Zhu, 1997). Dust from a steel plant not only reduced the yield of vegetables and fruit but also increased the heavy metal content of the crop and reduced its quality (Wu et al, 1990; Fu et al, 1996).

SO$_2$ in low concentrations can significantly affect the growth and yield of plant crops and also change plant sensitivity to other environmental stressors such as low temperatures, droughts, and pathogens and pests. In China, SO$_2$ alone was found in a study with field-based open-top chambers to be the main factor responsible for 5–8 per cent losses in yield in seven provinces (Feng et al, 1999). In India, the average yield loss in paddy due to exposure to cement dust was estimated to amount to almost 20 per cent (Varshney et al, 1997).

Ground level O$_3$ is a widespread phytotoxic pollutant that frequently exceeds the WHO air quality guideline value of 100 µg/m^3 and the critical levels AOT40 accumulated exposure over a threshold of 40 ppb (80 µg/m^3)[1] (WHO, 2000b; 2005) for agricultural crops across many parts of Asia. Emissions of O$_3$ precursors, in particular NO$_x$, have increased over the past decade due to rising energy demands resulting from urbanization, rapid increases in industrialization and motorized transport and increases in the use of nitrogenous fertilizers. A survey in Taiwan indicated that O$_3$ might be affecting sweet potato and spinach over a large area of the Taipei Basin (Sun, 1994). Crops of tropical fruits, vegetables and ornamental crops showed symptoms associated with O$_3$ injury while crops such as cucumber, muskmelon, guava and Indian jujube all showed signs of injury that could be attributable to the mixture of air pollutants (Lin and Yang, 1996). Wheat and soybean plants exposed to O$_3$ concentrations of less than 200 µg/m^3 showed a decrease of yield and biomass accumulation with increasing O$_3$ concentration in India. A study by Zheng et al (1998) indicated a potential for adverse growth effects of O$_3$ on several Chinese plant species (green pepper, rice, aubergine, cauliflower).

Since emissions of NO$_x$ are projected to rise until 2020, O$_3$ concentrations may also increase. This will result in an increasing frequency, magnitude and duration of ground-level O$_3$ pollution episodes across large areas of the South Asian region (Streets and Waldoff, 2000). As a consequence, there may be a substantial reduction in crop yields.

Experimental and observational studies conducted at locations across the South Asian region during the 1990s have indicated that current ambient O$_3$ levels can cause significant losses in agricultural productivity. O$_3$ exposure can result in yield reductions (up to 40 per cent for rice and approximately 50 per cent for wheat); visible injury reducing the economic value of leafy crops and the reduction of nutritional quality of grain and vegetables (Emberson et al, 2003). Given that current WHO critical level guidelines for Europe use a threshold of 40 ppb (80 µg/m^3) above which levels of O$_3$ are potentially damaging to agricultural crops, the threat to food production would seem to already exist and the implications for the future sustainability of agriculture may be serious. In addition to O$_3$ pollution, stresses to agricultural productivity may also arise from climate change (through altered precipitation and temperature patterns affecting drought and heat stress) and through the continued occurrence of the atmospheric brown cloud during the winter months.

The challenge for the Asian region is to provide sustainable increases in productivity to balance the reduced per capita area harvested.

Material assets and cultural heritage

In developing countries in Asia emissions of SO_2 in combination with nitrogen compounds, O_3 and PM pose a potential hazard to material assets and cultural heritage. Warm temperatures with high relative humidity and a high frequency of precipitation increase the potential corrosion damage, especially that caused by wet deposition. Corrosion is a process that causes damage to materials by physical, chemical or electrochemical reactions with the surrounding environment. Acidifying pollutants deposited on materials often dramatically increase deterioration rates. Materials affected include metal, painted surfaces, calcareous stones, rendering, polymer materials and paper (RAPIDC, 2004).

The effects on buildings, including loss of mechanical strength, leakage, failure of protective coatings or loss of details in carvings, provide some of the most striking evidence of damage related to fossil fuel use (Tiblad and Kucera, 1998). One key example of the effect of air pollution on cultural heritage is the Taj Mahal. In recent years, the impact of air pollution on the Taj Mahal has become a matter of grave concern (UNESCO, 2000). It has been recognized that pollutants from century-old industries and increased tourism and transportation in Agra and the Taj Trapezium Zone have had a corrosive impact on the buildling. Pollution has been partially eliminated through strict limits on the numbers of cars and buses allowed into the area near the monument, which is a first step in the conservation effort.

Economic impact of air pollution

The economic costs of urban air pollution run into hundreds of millions of US dollars. The majority of studies examining the economic impact of air pollution in Asia have focused on the costs attributed to the effects of air pollution on human health. The poor are particularly vulnerable to the effects of urban air pollution as they are more likely to be directly exposed to polluting emissions. They often suffer from poor nutrition which exacerbates any condition caused by exposure to high concentrations of air pollutants.

The World Bank estimated the costs related to PM_{10} for cities in the Philippines and Thailand. The estimated costs for Manila due to PM_{10} amounted to US$392 million, while the combined costs for the other three cities included in the study were only US$40 million in 2001 (World Bank, 2002). For Thailand, the costs attributed to PM_{10} for six cities totalled US$643.9 million in 2000, while for Bangkok they totalled US$424 million, equivalent to 0.5 per cent of the national GDP in 1999 (World Bank,

2002). Using a baseline scenario for the Clean Air for Europe (CAFE) project, the EU estimated that in 2000 the annual health damage due to air pollution ranged between 275.8 and 789.9 billion euro and would be reduced to between 188.8 and 608.9 billion euro in 2020 when the current EU legislation is implemented in the 25 member states (AEAT, 2005). The health benefits would amount to between 87.0 and 181.0 billion euro.

A 2001 ADB study estimated that the costs of premature mortality and morbidity in Jakarta attributed to PM_{10} were approximately US$181 million in 1998 (Syahril et al, 2002). In India, the estimated annual health damage of emissions before introduction of European standards for 25 cities ranged from a low of US$14 million to a high of US$191.6 million in 2002 (Mashelkar et al, 2002).

Health benefits were also estimated from studies as part of the USEPA Integrated Environmental Strategies programme for cities in China, India and South Korea. Different energy-related policies were analysed and compared to business-as-usual scenarios.

In Shanghai, the estimated health benefits derived from implementing various policies in controlling air pollution were in the range US$113–950 million in 2010 and US$327 million–2 billion in 2020 (Chen et al, 2001). Similarly, the estimates derived for Beijing amounted to US$270–760 million in 2010 and US$380 million–1 billion in 2020 (Department of Environmental Science and Engineering, Tsinghua University et al, 2005). In Hyderabad, India, the costs of impacts on health were estimated at US$200 million dollars per year in 2011 and US$1.2 billion dollars per year in 2021 (EPTRI, 2005). For Seoul, South Korea, the final assessment found that implementing integrated measures between 2000 and 2020 would result in cumulative health benefits of up to US$1.03 billion (1999), a significant figure when considering the low costs of the mitigation measures considered (Joh et al, 2001).

Only a small number of studies have investigated the monetary costs of air pollution impacts on agriculture in Asia. The Chinese Environmental Protection Agency estimated the economic losses due to damage caused by acid rain to forests and farmlands increased by five times from 1996 to 2000 to US$13.25 billion in 2000 (Shah et al, 2000). East Asian countries are considered to be on the cusp of substantial reductions in grain yield due to projected increases in O_3 concentrations (Wang and Mauzerall, 2004). It is estimated that East Asia lost 1–9 per cent of wheat, rice and corn yield and 23–27 per cent of soybean yield in 1990 due to O_3 concentrations. By 2020 it is estimated that grain loss due to damage by O_3 will increase to 2–16 per cent of wheat, rice and maize yield and 28–35 per cent of soybean yield. Compliance with O_3 air quality standards would increase yield by a value of US$2.6–27 billion in grain revenues (Wang and Mauzerall, 2004).

RESPONSE TO AIR POLLUTION

Clean air is recognized as a key component of a sustainable urban environment in international environmental agreements and increasingly in regional environmental declarations in Asia. Agenda 21, a result of the 1992 United Nations Conference on Environment and Development (UNCED), supports a number of environmental management principles on which governmental policies should be based, and promotes the adoption of national action programmes of urban air pollution (UNCED, 1992).

In 2002 the World Summit on Sustainable Development (WSSD) further recognized the problem of air pollution in Section IV 37 of its Plan of Implementation which requests states to enhance cooperation at the international, regional and national levels to reduce air pollution, including transboundary air pollution, acid deposition and stratospheric ozone depletion. The Plan also acknowledges the significant impact of air pollution on human health, in particular respiratory diseases and other health impacts resulting from air pollution, with special attention to women and children (WSSD, 2002).

Based on the recognition of the importance of environmental sustainability at the WSSD, the Association of Southeast Asian Nations (ASEAN) adopted the Framework for Environmentally Sustainable Cities. The aim of the Framework is to address environmental sustainability challenges in the areas of clean air, water and land. The area of clean air aims to achieve ASEAN's long-term goal of maintaining good ambient air quality to safeguard public health. It addresses emissions from both stationary sources as well as mobile sources and includes land-clearing activities (ASEAN, 2003).

The 2005 Declaration on Environment and Development at the Fifth Ministerial Conference on Environment and Development in Asia and the Pacific emphasized the need to improve environmental sustainability (UNESCAP, 2005). The Declaration calls upon the members and associate members of the Economic and Social Council for Asia Pacific to seek to improve environmental sustainability and address poverty by reducing the pressure of unsustainable economic growth on the environment and improving environmental performance.

The Ministerial Conference also adopted the Regional Implementation Plan for Sustainable Development in Asia and the Pacific Region, 2006–2010. The Plan provides a framework for action to address key constraints to sustainable development in the context of the needs and priorities of the region in the period following the WSSD (UNESCAP, 2005). In the same year the regional Environmentally Sustainable Transport (EST) forum adopted the Aichi Statement which promotes integrated transport policies, strategies and programmes incorporating air quality monitoring and assessment (UNCRD, 2005).

In 2001 the Clean Air Initiative for Asian Cities (CAI-Asia) was established to promote better AQM in Asian cities through the exchange of information, knowledge management, capacity building, policy dialogues and pilot projects. CAI-Asia has become the main regional initiative on AQM in Asia with local city networks in seven countries (CAI-Asia, 2006a). The now biannual Better Air Quality (BAQ) workshops are an important venue to review the state of the art in regional approaches to urban and transboundary air pollution in Asia. Future workshops aim to bring together senior government officials and representatives of key stakeholders from across Asia to achieve optimal air pollution abatement strategies (BAQ, 2006).

National and local government response

National and local governments have begun to develop AQM strategies to address the deterioration in urban air quality. However, the scope and effectiveness of such strategies vary widely between countries and cities. The aim of AQM is to maintain the quality of air that protects human health and welfare but also provides protection for animals, plants (crops, forests and natural vegetation), ecosystems, materials and aesthetics, such as natural levels of visibility. In order to achieve this goal, appropriate policies and strategies to prevent and control air pollution need to be developed and implemented. Some of these policies have been based on developments in Europe and North America.

It was once thought that by increasing the height of industrial chimney stacks, pollutants would be released much higher in the atmosphere and the environmental impacts would be reduced as a consequence of the dilution and dispersion of these pollutants. However, this approach to abating air pollution in the developed world has proved ineffective. More sophisticated strategies have involved 'end-of-pipe' abatement approaches, which involve the use of filters and scrubbers at the end of emission pipelines and chimney stacks to control the release of a particular air pollutant into the atmosphere, or prevention of the pollution altogether (Jackson, 1996).

Over the past 20 years, there has been a considerable expansion in air pollution regulation in Europe and North America. Table 1.7 provides an overview of the different types of environmental regulation available. Mandatory regulations and voluntary initiatives have demanded that businesses assume more responsibility for the environmental impacts which they create during the production process or in the final end product. However, voluntary initiatives, commonly used in the UK in the 1980s have not proved any more or less effective in achieving environmental objectives than the more adversarial and regulatory approach adopted by nations such as the US. US regulatory policy has been more ambitious and this has resulted in greater resistance from business. In contrast, it has been argued that because the UK

Table 1.7 *Types of environmental regulation (after WHO, 2000a)*

Type	Description	Examples
Command and control	Issue of licences, setting of standards, checking for compliance with standards, sanctions for non-compliance	Air pollution control regulations Government monitoring Emission standards Enforcement policies
Economic instruments	Use of pricing, subsidies, taxes and charges to change production and consumption patterns	Load-based emission charges Tradeable emission permits Differential taxes True cost pricing of resources
Co-regulation and voluntary initiatives	Adoption of rules, regulations and guidelines, negotiated within prescribed boundaries Voluntary adoption of environmental management measures	National registers of pollution emission inventories Environmental management systems
Self-regulation	Self-imposition of rules and guidelines and environmental audits by industry groups	Industry codes of practice Self-audit within industry groups Emission reduction targets
Education and information	Education and training Community right-to-know Corporate reporting programmes	Education, training and information programmes Pollution inventories Corporate sustainability reports

Source: WHO (2000a), Bradfield et al (1996)

approach demanded less, the demands were perceived as reasonable, and therefore industry was more likely to comply with environmental regulations (Vogel, 1986).

An alternative to the command and control approach is the use of market-based instruments such as emission charges, taxes on resource use, tax relief on new capital investments and penalties for the infringement of environmental regulations. For example, most European countries have encouraged the use of unleaded petrol and diesel by using taxes to make leaded petrol more expensive. Market-based instruments are seen to increase flexibility by allowing progressive shifts to less polluting fuels and by encouraging cost-effective abatement. The effectiveness of such instruments in achieving a particular environmental standard depends on the technical feasibility of removing pollutants, the elasticity of substitution among different fuels, the equitable availability of investment capital as well as general responsiveness of the market to price signals (Jackson, 1996).

Many Asian cities have developed some form of AQM system involving environmental regulation. However, the sophistication and completeness may not always be adequate to effectively address the problem of urban air pollution. To be effective, an AQM programme is dependent on a number of factors such as emission inventories and source apportionment, air quality monitoring networks, dispersion modelling, exposure and health and environmental impact assessment. A viable means is the development of emission and air quality standards together with a range of cost-effective pollution control measures and the legislative powers and resources to implement and enforce them (Elsom, 1996). In this context, public awareness and stakeholder participation are helpful to achieve the goal of better air quality.

The formulation and implementation of integrated AQM strategies requires the cooperation and coordination of national, regional and local authorities and the different groups of stakeholders that contribute to the problem. Many Asian countries have developed air quality prevention and control strategies including air quality standards for main pollutants as well as emission standards for power plants, certain industries and motor vehicles.

Options for reducing emissions range from measures which address fuel quality (including switching to cleaner fuels and improving the quality of fuels to reduce emissions), rationalization of fuel prices to provide incentives for efficient fuel use, adoption of technologies that reduce emissions at source, and energy efficiency measures that reduce emissions through reduction in the quantities of fuel used. Control technologies have been generally based on the modification of fuel or the combustion technique or removal of flue gases.

In most Asian countries lead in gasoline has been phased out while some countries have introduced cleaner fuels and cleaner vehicles and strengthened inspection and maintenance programmes. Hong Kong is an example of one city that used a creative taxing scheme to promote the use of unleaded fuel and low-sulphur diesel and was thus able to stop importing leaded fuel and the use of high-sulphur diesel. India has adopted a two-tier approach to adopting increasingly stringent fuel and vehicle requirements with the 11 largest cities tightening their standards at a faster pace than the rest of the country. India is also introducing one of the most stringent sets of requirements for its two- and three-wheelers and has thus significantly reduced the use of two-stroke engine motorcycles (PCFV, 2002).

Those countries that face severe air pollution problems have begun to adopt European standards for both fuel and automobile specifications and aim at being in parity with the Euro standards in the next decade (Bellagio Principles, 2001). Recently, China has taken significant steps to control motor vehicle emissions, phasing out leaded gasoline nationwide, replacing several hundred diesel buses with new natural gas buses, and within a few years converting most of Beijing's on-road car fleet to catalyst technologies with a combination of new vehicle emissions standards, retrofits and scrappage programmes. In December 2005 the State Council of China approved the implementation of State Phase III and IV (similar to Euro III and IV) vehicle emission standards in Beijing (Xin, 2005). In Japan, the national government has established a series of weight-class fuel economy standards that require an approximately 23 per cent improvement in the fuel economy of gasoline-fuelled light-duty vehicles by 2010 (Bellagio Principles, 2001).

Indonesia has adopted Euro II emission standards for all new vehicles and existing models by 2007. However, according to the Indonesian Ministry of Energy and Mineral Resources, Directorate General of Oil and Gas, corresponding fuel quality regulations are expected to be ready by 2006 (JAMA, 2006). Vietnam will adopt Euro II standards by 1 July 2007. The Philippines has implemented Euro II compliant diesel standards but has not yet instituted the adoption of Euro II vehicle emissions standards. The adoption of Euro II, for diesel- and gasoline-fuelled vehicles is planned by the end of 2006 (JAMA, 2006). Table 1.8 shows the level of sulphur in diesel while Table 1.9 shows the current and proposed vehicle emissions standards in Asia. A high sulphur level in diesel prevents the adoption of emission control technologies for vehicles and causes high sulphate concentrations. Most countries in the ASEAN region are examining the harmonization of fuel quality and vehicle emissions standards. Euro IV standards are currently envisaged to be implemented by ASEAN countries by 2010 with the exception of Indonesia and Vietnam (planned for 2012) (JAMA, 2006).

Other measures adopted to reduce motor vehicle pollution include diverting traffic away from heavily populated areas (e.g. by building ring roads around cities or restricting downtown traffic); converting high-use vehicles to cleaner fuels (e.g. converting buses to natural gas); improving vehicle maintenance; increasing the share of less polluting traffic modes; using more fuel-efficient vehicles; and installing catalytic control devices. Supply-side traffic management measures are aimed at reducing congestion (e.g. by improving road infrastructure). However, such measures rarely lead to significant overall emissions reductions because they may simply increase traffic flows.

Asian cities are increasingly implementing demand-side traffic management measures by prioritizing public transport users. Bangkok and Jakarta have a bus rapid transit (BRT) network to complement their existing public transportation system and reduce traffic congestion. Jakarta is in the process of opening two additional BRT corridors in 2006. Although the constructed BRT system has had a number of setbacks in terms of design and operations, the government is fully committed to improving the system and making it work. Other cities have also begun to examine the potential of BRT systems in improving public transportation, traffic flow and urban air quality.

Table 1.8 *Current and proposed sulphur levels in diesel in Asia, EU and US*

	1996	1997	1998	1999	2000	2001	2002	2003	2004	2005	2006	2007	2008	2009	2010	2011
Bangladesh	–	–	–	–	–	–	5000									
Cambodia	–	–	–	–	2000											
China (nationwide)	5000						2000			500 (Widely used)						
China (Beijing)	5000						2000		500	350						
Hong Kong		500					50					10 (Under consideration)				
India (nationwide)	5000				2500					500					350	
India (metropolitan)	5000				2500	500				350					50	
Indonesia	5000															
Japan	500				100					50/10						
Korea (Republic of)	500							430			30	10				
Malaysia	5000		3,000				500 (On Sale)			500					50	
Pakistan	10,000						5000									
Philippines	5000					2,000			500							
Singapore	3000		500								50					
Sri Lanka	10,000							5000	3000/500			500				
Taipei, China	3000			500			350			50						
Thailand	2500			500					350						50	
Vietnam	10,000							2,500			500					
EU					350					50/10			10			
US	500										15					

| ≥ 500 ppm |
| 51– 500 ppm |
| ≤ 50 ppm |

Source: CAI-Asia (2005)

Table 1.9 *Emission standards for new vehicles (light duty) in selected Asian countries*

	1995	1996	1997	1998	1999	2000	2001	2002	2003	2004	2005	2006	2007	2008	2009	2010
European Union	Euro I			Euro II				Euro III				Euro IV			Euro V	
Bangladesh (nationwide) gasoline												Euro II				
Bangladesh diesel												Euro I				
China (nationwide)							Euro I			Euro II			Euro III			Euro IV
China (Beijing and Shanghai)							Euro I		Euro II		Euro III					Euro IV[a]
Hong Kong, China	Euro I		Euro II				Euro III					Euro IV				
India (nationwide)							Euro I				Euro II					Euro III
India[b]					Euro I	Euro II					Euro III					
Indonesia											Euro II					
Malaysia			Euro I			Euro II								Euro IV[c]		
Nepal						Euro I										
Philippines									Euro I				Euro II		Euro IV	
Singapore (gasoline)	Euro I						Euro II									
Singapore (diesel)	Euro I						Euro II					Euro IV				
Sri Lanka									Euro I							
Taipei, China					US Tier I								US Tier II for diesel			
Thailand	Euro I						Euro II			Euro III					Euro IV	
Vietnam													Euro II			Euro III

Notes:

[a] Beijing only.

[b] New Delhi and other cities; Euro II introduced in Chennai, Kolkata, Mumbai in 2001; Euro II in Ahmedabad, Bangalore, Hyderabad, Khampur and Pune in 2003; Euro III introduced in Ahmedabad, Bangalore, Chennai, Hyderabad, Mumbai, Kolkata and New Delhi in 2005.

[c] Under discussion.

Source: CAI-Asia (2006b)

STRUCTURE OF THE BOOK

The book is divided into five chapters, including this introductory chapter which provides an overview of urban air pollution in Asia. Chapter 2 explains the methodological approach used to benchmark AQM in Asian cities. Chapter 3 provides an overview of AQM in 20 Asian cities. Chapter 4 discusses the different stages of AQM development in Asia and attempts to categorize cities according to their AQM capability. Finally, Chapter 5 outlines the key trends in AQM and key challenges faced by Asian cities in achieving better air quality. It concludes by providing general recommendations to improve AQM and achieve clean air in Asian cities.

NOTE

1 AOT40 means the sum of the difference between hourly concentrations greater than 80 $\mu g/m^3$ and 80 $\mu g/m^3$ over a given period using only the 1-hour values between 08:00 and 20:00, expressed in [($\mu g/m^3$)·hours] (EC, 2002).

REFERENCES

ADB/ASEAN (2001) 'Fire, Smoke and Haze: The ASEAN Response Strategy', Asian Development Bank, Association of Southeast Asian Nations, Manila, Philippines, www.aseansec.org/pdf/fsh.pdf accessed in February 2006

AEAT (2005) 'CAFÉ CBA: Baseline Analysis 2000 to 2020', AEA Technology Environment, Oxon, http://europa.eu.int/comm/environment/air/cafe/activities/pdf/cba_baseline_results2000_2020.pdf accessed in April 2006

Antle, J. M. and Heidebrink, G. (1995) 'Environment and Development: Theory and International Evidence', *Economic Development and Cultural Change*, vol 43, pp603–625

Agrawal, A. (2003) 'Air Pollution Impacts on Vegetation in India', in Emberson, L., Ashmore, M. and Murray, F. (eds) *Air Pollution Impacts on Crops and Forests: A Global Assessment*, London, Imperial College Press

Arrow, K., Bolin, B., Costanza, R., Dasgupta, P., Folke, C., Holling, C. S., Jansson, C.O., Levin, S., Mäler, K. G. Perrings, C. and Pimentel, D. (1995) 'Economic Growth, Carrying Capacity, and the Environment', *Science*, vol 268, pp520–521

ASEAN (2002) 'ASEAN Agreement on Transboundary Haze Pollution', Association of Southeast Asian Nations, Jakarta, www.aseansec.org/pdf/agr_haze.pdf accessed in February 2006

ASEAN (2003) 'Environmentally Sustainable Cities in ASEAN', Association of Southeast Asian Nations, Jakarta, www.aseansec.org/framework.htm accessed in February 2006

Bao, W. and Zhu, Z. (1997) 'Study of the Combination of Effects of Dust and SO_2 on Citrus Tree', *Journal of Agro-Environmental Protection*, vol 16, pp16–19 (in Chinese)

BAQ (2006) '5th Better Air Quality Workshop', Yogyakarta, 13–15 September 2006, www.cleanairnet.org/baq2006/1757/channel.html accessed in April 2006

Bellagio Principles (2001) 'Bellagio Memorandum on Motor Vehicle Policy. Principles for Vehicles and Fuels in Response to Global Environmental and Health Imperatives', The Energy Foundation, San Francisco, Beijing, www.hewlett.org/NR/rdonlyres/F7FFC376-1B4B-4C85-86D7-3E8AE55AE79B/0/Bellagio2001 memorandum.pdf accessed in February 2006

Bhattacharya, P. (2005) Personal Communication, West Bengal Pollution Control Board, Kolkata

Bi, X., Sheng, G., Peng, P. and Zhang, Z. (2002) 'Extractable Organic Matter in PM_{10} from LiWan district of Guangzhou city, PR China', *The Science of the Total Environment*, vol 300, pp213–228

BP (2004) 'Statistical Review of World Energy', British Petroleum, London, www.bp.com/liveassets/bp_internet/globalbp/STAGING/global_assets/downloads/S/statistical_review_of_world_energy_full_report_2004.pdf accessed in February 2006

Bradfield, P. J., Schulz, C. E. and Stone, M. J. (1996) 'Regulatory Approaches to Environmental Management', in Mulligan, D. R. (ed.) *Environmental Management in the Australian Minerals and Energy Industries*, Sydney, Australia, University of New South Wales Press

CAI-Asia (2005) 'Current and Proposed Sulfur Levels in Diesel – Asia, EU, US', Clean Air Initiative for Asian Cities, Manila, Philippines, www.cleanairnet.org/caiasia/1412/article-40711.html accessed in February 2006

CAI-Asia (2006a) 'Clean Air Initiative for Asian Cities', www.cleanairnet.org/caiasia/1412/channel.html accessed in April 2006

CAI-Asia (2006b) 'Emission Standards for New Vehicles (Light Duty)', Clean Air Initiative for Asian Cities, Manila Philippines, www.cleanairnet.org/caiasia/1412/articles-58969_new.pdf accessed in February 2006

CAI-Asia (2006c) 'Clean Air Initiative for Asian Cities Secretariat', www.cleanairnet.org/caiasia/1412/channel.html, accessed August 2006

Cao, J. J., Lee, S. C., Ho, K. F., Zhang, X. Y., Zou, S. C., Fung, K., Chow, J. C. and Watson, J. G. (2003) 'Characteristics of Carbonaceous Aerosol in Pearl River Delta Region, China, during 2001 Winter Period', *Atmospheric Environment*, vol 37, pp1451–1460

Chakraborti, D. (2003) 'Kolkata City: An Urban Air Pollution Perspective' in Whitelegg, J. and Haq, G. (eds) *An Earthscan Reader in World Transport Policy and Practice*, London, Earthscan

Chen, C., Fu, Q., Chen, M., Chen, B., Hong, C. and Kan, H. (2001) 'Benefit-Chapter 10: Economic Valuation of Health Outcomes Associated with Air Pollution under Various Energy Scenarios in Shanghai', Washington DC, United States Environmental Protection Agency, www.epa.gov/ies/documents/shanghai/full_report_chapters/ch10.pdf accessed in March 2006

Cofala, J., Amann, M., Gyarfas, F., Schoepp, W., Boudri, J. C., Hordijk, L., Kroeze, C., Junfeng, L., Lin, D., Panwar, D. S. and Gupta, S. (2004) 'Cost-effective Control of SO_2 Emissions in Asia', *Journal of Environmental Management*, vol 72, pp149–161

Cole, M. A. (2003) 'Development, Trade, and the Environment: How Robust is the Environmental Kuznets Curve?' *Environment and Development Economics*, vol 8, no 4, pp557–579

Cole, M. A., Rayner, A. J. and Bates, J. M. (1997) 'The Environmental Kuznets Curve: An Empirical Analysis', *Environment and Development Economics*, vol 2, pp401–416

Department of Environmental Science and Engineering, Tsinghua University, School of Public Health, Peking University, School of Public Health, Yale University, National Renewable Energy Laboratory (2005) 'Energy Options And Health Benefit Beijing Case Study', United States Environmental Protection Agency, Washington DC, www.epa.gov/ies/documents/general/Beijing%20Excutive%20summary_v3.0.pdf accessed in March 2006

Duggal V. K. and Pandey, G. K. (2002) 'Air Quality Management in Delhi', presented at BAQ 2002, Hong Kong, 16–18 December 2002, www.cse.polyu.edu.hk/~activi/BAQ2002/BAQ2002_files/Proceedings/CityFocus/cf-1Pandey.pdf accessed in February 2006

EANET (2005a) 'Acid Deposition Monitoring Network in East Asia', www.eanet.cc/eanet.html accessed in February 2006

EANET (2005b) 'EANET Publications: Data Report on the Acid Deposition in the East Asian Region 2000; 2001; 2002; 2003', Acid Deposition Monitoring Network in East Asia, www.eanet.cc/product.html accessed in February 2006

EEA (1999) 'DPSIR Conceptual Framework', European Environmental Agency, Copenhagen, Denmark, http://org.eea.eu.int/documents/brochure/brochure_reason.html accessed in February 2006

EEA (2003) 'Air Pollution in Europe 1990–2000', European Environment Agency, Copenhagen, http://reports.eea.eu.int/topicreport_2003_4/en/Topic_4_2003_web.pdf accessed in February 2006

EEA (2005) 'The European Environment. State and Outlook 2005', European Environmental Agency, Copenhagen, http://reports.eea.eu.int/state_of_environment_report_2005_1/en/tab_content_RLR accessed in April 2006

Elsom, D. (1996) *Smog Alert – Managing Urban Air Quality*, London, Earthscan

Emberson, L. D., Ashmore, M. R. and Murray, F. (eds) (2003) *Air Pollution Impacts on Crops and Forests: A Global Perspective*, London, Imperial College Press

Enerdata (2004) 'World Energy in 2003: Main Facts and Keys for Understanding', Enerdata, Paris, France, www.enerdata.fr/enerdatauk/ accessed in February 2006

EPTRI (2005) 'Hyderabad', Integrated Environment Strategies, Washington DC, United States Environmental Protection Agency, www.epa.gov/ies/documents/india/iesfinal_0405.pdf accessed in February 2006

Feng, Z., Cao, H. and Zhou, S. (1999) *Effects of Acid Deposition on Ecosystems and Recovery Study of Acid Deposition Damaged Forest*, Beijing, China Environmental Science Press (in Chinese)

Frankenberg, E., McKee, D. and Thomas, D. (2005) 'Health consequences of forest fires in Indonesia', *Demography* vol 42, pp109–129

Fu, J. (2004) Personal Communication, State Key Laboratory of Organic Geochemistry, Guangzhou Institute of Geochemistry, Guangzhou, Chinese Academy of Sciences

Fu, L., Meng, F., Liu, C., Chen, Q. and Ban, X. (1996) 'The Effects of Cement Dust on Soil and Crops', *Journal of Agro-Environmental Protection*, vol 15, pp221–224 (in Chinese)

Gangadharan, L. and Valenzuela, M. R. (2001) 'Interrelationships between Income, Health and the Environment: Extending the Environmental Kuznets Curve Hypothesis' *Ecological Economics*, vol 36, no 3, pp513–531

Goklany, I. M. (1999) *Clearing the Air: The Real Story of the War on Air Pollution*, Washington DC, Cato Institute

Gouveia, N. and Fletcher, F. (2000) 'Time Series Analysis of Air Pollution and Mortality: Effects by Cause, Age and Socioeconomic Status', *Journal of Epidemiology and Community Health*, vol 54, pp750–755

HEI (2003) *Revised Analyses of Time-series Studies of Air Pollution and Health*, special report, Boston, MA, Health Effects Institute

HEI (2004) *Health Effects of Outdoor Air Pollution in Developing Countries of Asia: A Literature Review*, HEI Special Report 15, Boston, MA, Health Effects Institute

Hettige, H., Robert, L. E. B. and Wheeler, D. (1992) 'The Toxic Intensity of Industrial Production: Global Patterns, Trends and Trade Policy' *American Economic Review*, vol 82, pp478–481

Hien, P. D., Bac, V. T. and Thinh, N. T. H. (2004) 'PMF Receptor Modelling of Fine and Coarse PM_{10} in Air Masses Governing Monsoon Conditions in Hanoi, Northern Vietnam', *Atmospheric Environment*, vol 38, pp189–201

Hill, R. J. and Magnani, E. (2002) 'An Exploration of the Conceptual and Empirical Basis of the Environmental Kuznets Curve', *Australian Economic Papers*, vol 41, pp239–254

IEA (2004) 'International Energy Outlook 2004', Energy Information Administration, Washington DC, USA, www.eia.doe.gov/oiaf/archive/ieo04/index.html accessed in February 2006

IGES (2004) *Urban Energy Use and Greenhouse Gas Emissions in Asian Mega-Cities*, Urban Environmental Management Project, Kitakyushu, Japan, Institute for Global Environmental Strategies

Jackson, T. (1996) *Material Concerns: Pollution, Profit and Quality of Life*, London, Routledge

JAMA (2006) *Japan Automobile Manufacturers Association Newsletter*, vol 18, February 2006, www.jama-english.jp/asia/news/vol18.pdf accessed in February 2006

Joh, S., Nam, Y, Shim, S. Sung, J. and Shin, Y. (2001) *Ancillary Benefits Due to Greenhouse Gas Mitigation, 2000–2020 – The International Co-Control Analysis Program for Korea*, Washington DC, United States Environmental Protection Agency, www.epa.gov/ies/documents/korea/finalreport/fullreport_korea.pdf accessed in March 2006

Katsouyanni, K., Touloumi, G., Spix, C., Schwartz, J., Balducci, F., Medina, S., Rossi, G., Wojtyniak, B., Sunyer, J., Bachnova, L., Schouten, J. P., Ponka, A. and Anderson, H. R. (1997) 'Short Term Effects of Ambient Sulphur Dioxide and Particulate Matter in 12 European Cities: Results from Time Series Data from the APHEA Project', *British Medical Journal*, vol 314, pp1658–1663

Katsouyanni, K., Touloumi, G., Samoli, E., Gryparis, A., Le, T. A., Monopolis, Y., Rossi, G., Zmirou, D., Ballester, F., Boumghar, A., Anderson, H. R., Wojtyniak, B., Paldy, A., Braunstein, R., Pekkanen, J., Schindler, C. and Schwartz, J. (2001) 'Confounding and Effect Modification in the Short-term Effects of Ambient Particles on Total Mortality: Results from 29 European cities within the APHEA2 project', *Epidemiology*, vol 12, pp521–531

Lam, K. S., Wang, T. J., Wu, C. L. and Li, Y. S. (2005) 'Study on an Ozone Episode in Hot Season in Hong Kong and Transboundary Air Pollution over Pearl River Delta Region of China', *Atmospheric Environment*, vol 39, pp1967–1977

Lenzen, M., Wier, M., Cohen, C., Hayami, H., Pachauri, S. and Schaeffer, R. (2006) 'A Comparative Multivariate Analysis of Household Energy Requirements in Australia, Brazil, Denmark, India and Japan', *Energy*, vol 31, pp181–207

Li, J., Guttikunda, S. K., Carmichael, G. R., Streets, D. G., Chang, Y. S. and Fung, V. (2004) 'Quantifying the Human Health Benefits of Curbing Air Pollution in Shanghai' *Journal of Environmental Management*, vol 70, pp49–62

Lin, C. C. and Yang, S. H. (1996) 'The Effect of Atmospheric Quantities on Horticultural Crops in Southern Taiwan', *Chinese Journal of Agrometeorology*, vol 4, no 4, pp183–196 (in Chinese)

Lindmark, M. (2002) 'An EKC-Pattern in Historical Perspective: Carbon Dioxide Emissions, Technology, Fuel Prices, and Growth in Sweden 1870–1997', *Ecological Economics*, vol 42, no 2, pp333–347

LRTAP (2005) 'Task Force on the Hemispheric Transport of Air Pollution. Long-range Transboundary Air Pollution', United Nations Economic Commission for Europe, Geneva, www.htap.org/index.htm accessed in February 2006

Marland, G., Boden, T. A. and Andres, R. J. (2005) 'Online Trends: A Compendium of Data on Global Change', http://cdiac.esd.ornl.gov/trends/emis/em_cont.htm accessed in February 2005

Martins. M. C. H., Fatigati, F. L., Ve´spoli, T. C., Martins, L. C., Pereira, L. A. A., Martins, M. A., Saldiva, P. H. N. and Braga, A. L. F. (2004) 'Influence of Socio-economic Conditions on Air Pollution: Adverse Health Effects in Elderly People: An Analysis of Six Regions in Sao Paulo, Brazil', *Journal of Epidemiology and Community Health*, vol 58, pp41–46

Mashelkar R. A., Biswas D. K., Krishnan, N. R., Mathur, O. P., Natarajan R., Niyati, K. P., Shukla, P. R., Singh, D. V. and Singhal, S. (2002) *Report of the Expert Committee on Auto Fuel Policy. Ministry of Petroleum and Natural Gas*, New Delhi, India, Government of India

McGranahan, G., Jacobi, P., Songsore, J., Surjadi, C. and Kjellen, M. (2001) *The Citizens at Risk: From Urban Sanitation to Sustainable Cities*, London, Earthscan

Min, B. S. (2001) 'Regional Cooperation for Control of Transboundary Air Pollution in East Asia' *Journal of Asian Economics*, vol 12, pp137–153

MoE (Ministry of Environment) Japan (1998) 'State of the Global Environment at a Glance: Acid Deposition', Tokyo, Ministry of Environment, www.env.go.jp/en/topic/acid/acid_situ.html accessed in February 2006

Munasinghe, M. (1999) 'Is Environmental Degradation an Inevitable Consequence of Economic Growth? Tunnelling through the Environmental Kuznets Curve', *Ecological Economics*, vol 29, no 1, pp89–109

NEERI (2004) *Long-term Study of Air Quality Trends in Major Urban Centres*, Mumbai, Zonal Centre, National Environmental Engineering Research Institute

PCFV (2002) 'Partners for Cleaner Fuels and Vehicles' Steering Committee Meeting', United Nations Headquarters, 14–15 November 2002, www.unep.org/pcfv/main/main.htm accessed in February 2006

RAPIDC (2004) 'Regional Air Pollution in Developing Countries (RAPIDC) Programme', Stockholm, Sweden, Stockholm Environment Institute, www.rapidc.org/ accessed in February 2006

Samet, J. M., Zeger, S. L., Dominici, F., Curriero, F., Coursac, I., Dockery, D. W., Schwartz, J. and Zanobetti, A. (2000) *The National Morbidity, Mortality, and Air Pollution Study, Part II: Morbidity and Mortality from Air Pollution in the United States*, Research Report 94, Cambridge MA, Health Effects Institute

Sastry, N. (2002) 'Forest Fires, Air Pollution, and Mortality in Southeast Asia', *Demography*, vol 39, pp1–23

Schwela, D. (2000) 'Air Pollution and Health in Urban Areas', *Reviews on Environmental Health*, vol 15, pp13–42

Seip, H. M., Aagard, P., Angel, V., Eilertsen, O., Larssen, T., Lydersen, E., Mulder, J., Muniz, I. P., Semb, A., Tang, D., Vogt, R. D., Xiao, J., Xiong, J., Zhaio, D. and Kong, G. (1999) 'Acidification in China: Assessment Based on Studies at Forested Sites from Chongqing to Guangzhou', *Ambio*, vol 28, pp522–528

Seldon, T. M. and Sorg, D. (1994) 'Environmental Quality and Development: Is There a Kuznets Curve for Air Pollution Emissions?', *Journal of Environmental Economics and Management*, vol 27, pp147–162

Shafik, N. (1994) 'Economic Development and Environmental Quality: An Econometric Analysis', *Oxford Economic Papers*, vol 46, pp757–777

Shah J., Nagpal, T., Johnson, T., Amann, M., Carmichael, G., Foell, W., Green, C., Hettelingh, J., Hordijk, L., Li, J., Peng, C., Pu, Y., Ramankutty, R. and Streets, D. (2000) 'Integrated Analysis for Acid Rain in Asia. Policy Implications and Results of RAINS-Asia Model', *Annual Review of Energy and Environment*, vol 25, pp339–375

Sheu, B. H. and Liu, C. P. (2003) 'Air Pollution Impacts on Vegetation in Taiwan', in Emberson, L., Ashmore, M. and Murray, F. (eds) *Air Pollution Impacts on Crops and Forests. A Global Assessment*, London, Imperial College Press

Sinton, J. E. and Fridley, D. (2000) 'What Goes Up: Recent Trends in China's Energy Consumption', *Energy Policy*, 28, pp671–687

Sinton, J. E., Stern, R. E., Aden, N. T. and Levine, M. D. (2005) 'Evaluation of China's Energy Strategy Options', The China Sustainable Energy Program, China Energy Group, Berkeley, US, http://china.lbl.gov/publications/nesp.pdf accessed in February 2006

Smith, K. R. (1997) 'Development, Health, and the Environment Risk Transition' in Shahi, G., Levy, B. S., Binger, A., Kjellstrom, T. and Lawerance, R. (eds) *International Perspectives in Environment and Development, and Health: Toward a Sustainable World*, A Collaborative Initiative of the World Health Organization, the United Nations Development Programme, and the Rockefeller Foundation, New York, Springer

Smith, K. R. and Ezzati, M. (2005) 'How Environmental Health Risks Change with Development: The Environmental Risk and Epidemiologic Transitions Revisited', *Annual Review of Energy and Resources*, vol 30, pp291–333

Stern, R. E. (2003) 'Hong Kong Haze: Air Pollution as a Social Class Issue', *Asian Survey*, vol 43, no 5, p780, p21

Streets, D. G. and Waldoff, S. T. (2000) 'Present and Future Emissions of Air Pollutants in China: SO_2, NO_x, and CO', *Atmospheric Environment*, vol 34, pp363–374

Streets, D. G., Gupta, S., Waldhoff, S. T., Wang, M. Q., Bond, T. C. and Yiyun, B. (2001) 'Black Carbon Emissions in China', *Atmospheric Environment*, vol 35, pp4281–4296

Streets, D. G., Bond, T. C., Carmichael, G. R, Fernandes, S., Fu, Q., He, D., Klimont, Z., Nelson, S. M., Tsai, N. Y., Wang, M. Q., Woo, J. H. and Yarber, K. F. (2003a) 'A Year-2000 Inventory of Gaseous and Primary Aerosol Emissions in Asia to Support TRACE-P Modeling and Analysis,', www.cgrer.uiowa.edu/woo/summary/summary_2000_final.htm accessed in February 2006

Streets, D. G., Bond, T. C., Carmichael G. R., Fernandes, S. D. and Fu, Q. (2003b) 'An Inventory of Gaseous and Primary Aerosol Emissions in Asia in the Year 2000', *Journal of Geophysical Research*, vol 108, 8819, doi:10.1029/2002JD003093

Sun, E. J. (1994) 'Ozone Injury to Leafy Sweet Potato and Spinach in Northern Taiwan', *Botanical Bulletin of Academia Sinica*, vol 35, pp165–170 (in Chinese)

Sun, Y., Zhuang, G., Wang, Y., Han, L., Guo, J., Dan, M., Zhang, W., Wang, Z. and Hao, Z. (2004) 'The Air-borne Particulate Pollution in Beijing – Concentration, Composition, Distribution and Sources', *Atmospheric Environment*, vol 38, pp5991–6004

Syahril, S., Resosudarmo, B. and Tomo, H. S. (2002) 'Study on Air Quality, Future Trends, Health Impacts, Economic Value and Policy Options, Jakarta, Indonesia', Asian Development Bank Regional Technical Assistance 5937: Reducing Vehicle Emissions in Asia, Manila, Philippines, www.cleanairnet.org/caiasia/1412/articles-36531_recurso_1.pdf accessed March 2006

Tiblad, J. and Kucera, V. (1998) 'Corrosion of Man-made Materials', in Kuylenstierna, J. and Hicks, K. (eds) *Regional Air Pollution in Developing Countries*, Background Document for Policy Dialogue, Bangkok, March 1998, Stockholm, Sweden, Stockholm Environment Institute

Torras, M. and Boyce, J. K. (1998) 'Income, Inequality, and Pollution: A Reassessment of the Environmental Kuznets Curve', *Ecological Economics*, vol 25, no 2, pp147–160

UN (2004) 'World Urbanization Prospects: The 2003 Revision', ST/ESA/SER. A/237, United Nations, Department of Economic and Social Affairs, Population Division, New York, www.un.org/esa/population/publications/wup2003 accessed in February 2006

UNCED (1992) Agenda *21: Programme of Action for Sustainable Development*, United Nations Conference on Environment and Development, 3–14 June 1992, Rio de Janiero, www.un.org/esa/sustdev/documents/agenda21/index.htm accessed in February 2006

UNCRD (2005) 'The Aichi Statement on Environmentally Sustainable Transport', United Nations Centre for Regional Development, Nagoya, Japan, www.uncrd. or.jp/env/est/regional_est_forum/first_regional_est_forum_top.htm accessed in February 2006

UNECE (2004) 'Decision 2004/4 Concerning the Establishment of a Task Force on the Hemispheric Transport of Air Pollution', United Nations Economic Commission for Europe, Geneva, www.unece.org/env/tfhtap/TFonHT.Decision.2004.4.pdf accessed in February 2006

UNEP (2002) *Global Environment Outlook 3*, United Nations Environment Programme, London, Earthscan, www.unep.org/geo/geo3/ accessed in February 2006

UNEP (2004) 'North East Asian Dust and Sand Storms Growing in Scale and Intensity', Global Resource Information Database, United Nations Environment Programme, Nairobi, www.unep.org/Documents.Multilingual/Default.asp?DocumentID=388&ArticleID=4460&l=en accessed in February 2006

UNEP (2005) *GEO Year Book 2004/5: An Overview of Our Changing Environment*, UNEP/GC.23/INF/2, Nairobi, United Nations Environment Programme, www.unep.org/geo/yearbook/ accessed in February 2006

UNEP/WHO (1992) *Urban Air Pollution in Megacities of the World*, United Nations Environment Programme/World Health Organization, Oxford, Blackwell

UNESCAP (2005) 'The Regional Implementation Plan for Sustainable Development in Asia and the Pacific Region, 2006-2010', Bangkok, Thailand, www.unescap.org/mced/documents/presession/english/SOMCED5_5E_RIP.pdf accessed in February 2006

UNESCO (2000) 'Toxins and the Taj', *The Courier*, July/August 2000, www.unesco.org/courier/2000_07/uk/signe.htm accessed in February 2006

UN-HABITAT (2004) *The State of the World's Cities 2004/5: Globalization and the Urban Culture*, London, Earthscan, www.unhabitat.org/ accessed in February 2006

Varshney, C. K., Agrawal, M., Ahmad, K. J., Dubey, P. S. and Raza, S. H. (1997) *Effect of Air Pollution on Indian Crop Plants*, Final Report, UK ODA Project, Imperial College of Science, Technology and Medicines,

Vingarzan, R. (2004) 'A Review of Surface Ozone Background Levels and Trends', *Atmospheric Environment*, vol 38, pp3431–3442

Vogel, D. (1986) *National Styles of Regulation: Environmental Policy in Great Britain and the United States*, Ithaca, Cornell University Press

Wahid, A. (2003) 'Air Pollution Impacts on Vegetation in Pakistan', in Emberson, L., Ashmore, M. and Murray, F. (eds) *Air Pollution Impacts on Crops and Forests. A Global Assessment*, London, Imperial College Press

Wang, B., Peng, Z., Zhang, X., Xu, Y., Wang, H., Allen, G., Wang, L. and Xu, X. (1999) 'Particulate Matter, Sulfur Dioxide, and Pulmonary Function in Never-smoking Adults in Chongqing, China', *International Journal of Occupational and Environmental Health*, vol 5, pp14–19

Wang, X., Mauzerall, D. L., Hu, Y., Russell, A. G., Larson, E. D., Woo, J. H., Streets, D. G. and Guenther, A. (2005) 'A High Resolution Inventory for Eastern China in

2000 and Three Scenarios for 2020', *Atmospheric Environment*, vol 39, pp5917–5933

Wang, X. and Mauzerall, D. (2004) 'Characterising Distributions of Surface Ozone and its Impact on Grain Production in China, Japan, South Korea 1990 and 2020', *Atmospheric Environment*, vol 38, pp4383–4402

WBCSD (2004) 'Mobility 2030: Meeting the Challenge', World Business Council on Sustainable Development, Geneva, Switzerland, www.wbcsd.org/web/publications/mobility/mobility-full.pdf accessed in February 2006

WHO (2000a) *Guidelines for Air Quality*, WHO/SDE/OEH/00.02, Geneva, World Health Organization

WHO (2000b) *Air Quality Guidelines for Europe*, WHO Regional Publications, European Series, World Health Organization, Regional Office for Europe, Copenhagen, www.who.dk accessed in February 2006

WHO (2001) 'Air Quality and Health, Air Management Information System (AMIS 3.0)', (CD-ROM), Geneva, World Health Organization

WHO (2002) *The World Health Report 2002: Reducing Risks, Promoting Healthy Life*, Geneva, World Health Organization

WHO (2003) *Health Aspects of Air Pollution with Particulate Matter, Ozone and Nitrogen Dioxide*, Report of a WHO Working Group, Bonn, 13–15 January 2003, www.who.dk/document/e79097.pdf accessed in February 2006.

WHO (2004) *Meta-analysis of Time-Series Studies and Panel Studies of Particulate Matter (PM) and Ozone (O_3)*, Report of a WHO Task Group, World Health Organization, Copenhagen, WHO Regional Office for Europe, www.euro.who.int/document/e82792.pdf accessed in February 2006

WHO (2005) 'WHO Air Quality Guidelines Global Update 2005', Report of a working group meeting, Bonn, Germany, 18–20 October 2005, World Health Organization, Geneva, www.cepis.ops-oms.org/bvsea/fulltext/guidelines05.pdf accessed in April 2006

WHO/UNEP/WMO (1999) *Health Guidelines for Vegetation Fire Events*, Geneva, World Health Organization, Nairobi, United Nations Environment Programme, Geneva, World Meteorological Organization, www.who.int/mediacentre/factsheets/fs254/en/ accessed in February 2006

World Bank (1992) *World Development Report 1992*, Washington DC, The World Bank

World Bank (2002) '*Thailand Environment Monitor 2002*', Washington DC, World Bank, www.worldbank.or.th/WBSITE/EXTERNAL/COUNTRIES/EASTASIA PACIFICEXT/THAILANDEXTN/0,,contentMDK:20206650~pagePK: 141137~piPK:217854~theSitePK:333296,00.html accessed in March 2006

World Bank/ESMAP (2004) 'What is Causing Particulate Air Pollution? Evidence from Delhi, Kolkata, and Mumbai', *South Asia Urban Air Quality Management*

Briefing Note No. 14, Washington DC, The World Bank, http://lnweb18.worldbank.org/sar/sa.nsf/Attachments/Briefing14/$File/Briefing_Note_No_14.pdf accessed in February 2006

WSSD (2002) *World Summit on Sustainable Development – Plan of Implementation*, Johannesburg, South Africa, www.johannesburgsummit.org/ accessed in February 2006

Wu, D., Qing, C. and Gao, S. (1990) 'Study of the Effects of Dust Emitted from a Steel Plant on Soil-Vegetable System', *Journal of Agro-Environmental Protection*, vol 9, pp13–16 (in Chinese)

Xin, Y. (2005) 'Beijing Formally Launched Phase III (Euro III) Emission Standards', English translation, www.cleanairnet.org/caiasia/1412/article-70217.html accessed in February 2006

Yandle, B., Bhattarai, M. and Vijayaraghavan, M. (2004) 'Environmental Kuznets Curves: A Review of Findings, Methods and Policy Implications', Property and Environment Research Center (PERC), Research Study 1–2 April 2004

Zheng J., Tan, M., Shibata, Y., Tanaka, A., Li, Y., Zhang, G., Zhang, Y. and Shan, Z. (2004) 'Characteristics of Lead Isotope Ratios and Elemental Concentrations in PM_{10} Fraction of Airborne Particulate Matter in Shanghai after the Phase-out of Leaded Gasoline', *Atmospheric Environment*, vol 38, pp1191–1200

Zheng, Y. and Shimizu, H. (2003) 'Air Pollution Impacts on Vegetation in China', in Emberson, L., Ashmore, M. and Murray, F. (eds) *Air Pollution Impacts on Crops and Forests. A Global Assessment*, London, Imperial College Press,

Zheng, Y., Stevenson, K. J., Barrowcliffe, R., Chen, S., Wang, H. and Barnes, J. D. (1998) 'Ozone Levels in Chongqing: A Potential Threat to Crop Plants Commonly Grown in the Region?', *Environmental Pollution*, vol 99, pp299–308

two

Air Quality Management Capability in Asian Cities

INTRODUCTION

The majority of Asian cities are currently experiencing similar challenges in addressing urban air pollution and the exceedance of international air quality guidelines and standards (EU, 1999; 2000; 2002; USEPA, 1997; 2006; WHO, 2000; 2005). However, some advanced countries and cities (e.g. Hong Kong and Singapore) have successfully developed AQM capabilities and have seen improvement in air quality.

Considerable interest among policy makers exists for international comparisons of urban air pollution trends and policy measures. Comparative analyses of urban air pollution (e.g. UNEP/WHO, 1992; WHO/UNEP/MARC, 1996) and benchmarking studies (e.g. BEST, 2000; 2003; OECD, 2002) have demonstrated that learning from current practice can contribute to a better understanding and more effective implementation of environmental policies. By understanding the current stage of a city's development, additional action can be outlined to effectively reduce pollutant emissions and achieve better air quality (Gudmundsson, 2003).

This chapter outlines the methodological approach undertaken to provide a systematic assessment of AQM in 20 Asian cities. It provides an overview of the current state of AQM in these cities, which can assist in determining the city's stage in its development of air pollution problems.

SELECTION OF ASIAN CITIES

The 20 cities examined here differ widely in topography, size, complexity and level of economic development. They include megacities as well as smaller cities which are members of the CAI-Asia city network. The range of cities allows a comparative assessment of the different challenges encountered at the various stages of economic development. The choice of cities was partly dependent on whether it was possible to obtain information. This introduces an unavoidable bias in the selection as those cities unable or unwilling to provide information may be the cities with the most limited capabilities in AQM.

The cities covered here are: Bangkok, Beijing, Busan, Colombo, Dhaka, Hanoi, Ho Chi Minh City, Hong Kong, Jakarta, Kathmandu, Kolkata, Metro Manila, Mumbai, New Delhi, Seoul, Shanghai, Singapore, Surabaya, Taipei and Tokyo (see Figure 2.1 and Table 2.1).

METHODOLOGICAL APPROACH

The approach taken to provide an assessment of AQM in 20 Asian cities consists of three components:

1   Assessment of city authorities' AQM capabilities using a questionnaire survey completed by city authorities;
2   Compilation of information on current policy and practice for key aspects of AQM in the form of a City Profile; and

Figure 2.1 *Cities covered in the study*

Table 2.1 *Data on 20 Asian cities*

City	Coordinates[a]	Climate[a]	UN estimate (2003) (million)[b]	UN projection[b] (2015) (million)
Bangkok	13° 44'N, 100° 33'E	Tropical savannah	6.49	7.46
Beijing	39° 52'N, 116° 27'E	Continental warm summer	10.85	11.06
Busan	35° 09'N, 129° 03'E	Humid sub-tropical	3.58	3.40
Colombo	6° 56'N, 79° 50'E	Tropical rain forest	—	—
Dhaka	23° 43'N, 90° 24'E	Tropical rain forest	11.56	17.91
Hanoi	21° 01'N, 105° 51'E	Humid sub-tropical	3.98	5.28
Ho Chi Minh City	10° 45'N, 106° 42'E	Tropical savannah	4.85	6.31
Hong Kong	22° 16'N, 114° 11'E	Sub-tropical	7.05	7.87
Jakarta	6° 07'S, 106° 48'E	Tropical rain forest	12.30	17.50
Kathmandu	27° 42'N, 85° 18'E	Humid sub-tropical		
Kolkata	22° 32'N, 88° 21'E	Tropical savannah	13.81	16.80
Metro Manila	14° 34'N, 120° 59'E	Tropical rain forest	10.35	12.64
Mumbai	18° 56'N, 72° 51'E	Tropical savannah	17.43	22.65
New Delhi	28° 38'N, 77° 17'E	Dry steppe	14.15	20.95
Seoul	37° 32'N, 127° 00'E	Continental warm summer	9.71	9.21
Shanghai	31° 14'N, 121° 30'E	Humid sub-tropical	12.76	12.67
Singapore	1° 19'N, 103° 51'E	Tropical rain forest	4.25	4.71
Surabaya	7° 13'S, 112° 45'E	Tropical rain forest	2.62	3.45
Taipei	25° 04'N, 121° 33'E	Humid sub-tropical	2.51	2.45
Tokyo	35° 41'N, 139° 48'E	Humid sub-tropical	35.00	36.21

Notes:
[a] The Times (2003).
[b] UN (2004).

3 Comparative analysis based on current policy and practice and some general recommendations to improve AQM in Asian cities.

Air quality management capability

In order to assess the AQM capabilities of the 20 Asian cities the WHO/UNEP/MARC AQM capability index was used. The WHO/UNEP/MARC study (1996) developed an AQM capability index with indicators to assess each component of capability. This AQM capability index questionnaire survey was applied to 20 major cities throughout the world and 64 major European cities (WHO/UNEP/MARC, 1996; EEA, 1998). The index provides a useful tool to identify deficiencies and to allow comparisons between cities at different stages of economic development. Four sets of indicators (indices) to represent the key components of AQM capability were used:

1 **Air quality measurement capacity index**
   Assesses the ambient air monitoring taking place in a city and the accuracy, precision and representativeness of the data collected.
2 **Data assessment and availability index**
   Assesses how air quality data are processed to determine their value and how they are used to provide information in a decision-relevant format. It also assesses the extent to which there is access to air quality information and data through different media.
3 **Emission estimates index**
   Assesses emission inventories undertaken to determine the extent to which decision-relevant information is available about the sources of pollution in the city.
4 **Management enabling capabilities index**
   Assesses the administrative and legislative framework through which emission control strategies are introduced and implemented to manage air quality.

Each of the four component indices consists of a number of indicators, which are designed to determine whether a city has useful capacity with respect to a particular element of management capability. Figure 2.2 presents the different AQM indices and their constituent indicators.

   Using the AQM capability index, city authorities in the 20 Asian cities were asked to complete a questionnaire with Yes or No answers for a number of the component indicators (see Annex I). In order to effectively manage urban air pollution, a city authority requires capabilities for each component of the AQM system. (WHO/UNEP/MARC, 1996). An assessment of AQM capabilities enables the identification of the strengths and weakness of a city's current capabilities.

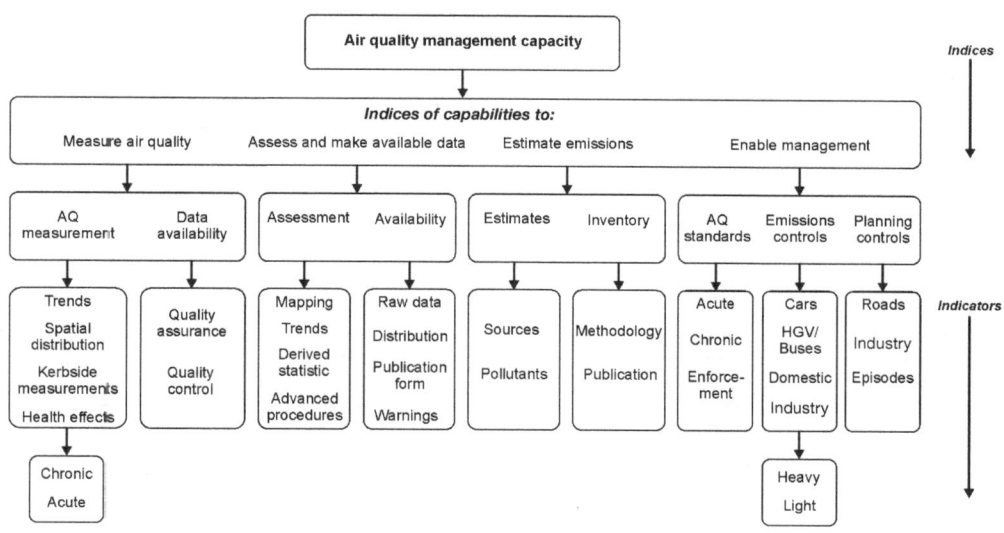

Figure 2.2 *Hierarchy of AQM capability indicators*

Source: WHO/UNEP/MARC (1996)

Representatives from city authorities, some of whom are members of the Clean Air Initiative for Asian Cities network (CAI-Asia, 2006), provided the information for this assessment. Questionnaire responses were reviewed by external experts and have been cross-checked, where possible, with other available information to ensure that the data presented are correct and that the questions have been fully understood. The survey was undertaken in the period 2002–2004 and additional clarification and updates were made until May 2005.

Each question was allocated a score. The total number of indicator points available for each component index is 25. The more questions answered positively the greater the management capability of that city. The original questionnaire was modified taking into consideration technological developments and the situation unique to Asia. The revised questionnaire and the steps of revision are presented in Annex II.

Table 2.2 presents the grouping of scores obtained into five bands of AQM capabilities. Each component index has also been combined to provide an overall assessment of the AQM capability. Each index has a maximum score of 25 indicator points; by adding these together an overall assessment of capability is obtained with a maximum score of 100. An equal weighting of each component index has been given to the overall assessment. The same five bands used for the component index are used for the overall capability index. The bands describe a range of values comprising a particular level of capability (WHO/UNEP/MARC, 1996).

Table 2.2 *Bandings for the component and overall capability indices*

Effectiveness of capability	Component index score	Overall capability index score
Minimal	0 – ≤5	0 – ≤20
Limited	>5 – ≤10	>20 – ≤40
Moderate	>10 – ≤15	>40 – ≤60
Good	>15 – ≤20	>60 – ≤80
Excellent	> 20 – ≤25	>80 –≤100

Note: ≤ smaller than or equal to; > larger than.

Source: WHO/UNEP/MARC (1996)

The capability index provides quantitative information which can be used to examine the relationship between different components of AQM. The index scores obtained from the answers given in the questionnaire provide an indication of the capabilities of Asian cities with respect to the four main components: (i) air quality management, (ii) data availability and access, (iii) estimation of emissions and (iv) presence of management enabling capabilities. The index scores identify the areas where the city could develop more effective capacity. It allows a comparison with the level of capabilities of other cities of similar and different levels of development.

An external review process was conducted and comments received were used to revise and update the survey information.

City profile

In addition to the AQM capability questionnaire, information was collated by local experts on AQM for each city in the form of a city profile. The local experts provided information for each city profile, which covered the following seven areas:

1 general information
2 air quality management
3 emissions
4 air quality monitoring
5 air quality situation
6 health and environmental impacts
7 conclusions.

The information in the city profile was provided by the cities to give an overview of AQM and provide a context for the results of the capability questionnaire and air quality trends. This information was reviewed by at least two independent experts based in the city or country. The city profiles were also used to clarify questions or discrepancies within the questionnaire. A summary of each city profile is presented in Chapter 3.

Therefore, the overall final classification of cities in the questionnaire survey was influenced positively or negatively by the profiles. For most cities, this procedure led to a change in capabilities scores of less than 10 per cent. In a few cases, for example Colombo, the scoring was substantially changed by the city profile, increasing the score by approximately 30 per cent.

AIR QUALITY MANAGEMENT CAPABILITY

The following sections examine the capacity of each city with respect to different elements of AQM capability. The AQM capability index has been used to provide a quantitative means of assessment and to determine the current state of AQM in Asian cities. Four indices on AQM capabilities have been calculated:

1   air quality measurement capacity
2   data assessment and availability
3   emission estimates
4   management enabling capabilities.

This AQM index and the city profile provide a benchmark against which future progress can be assessed and will serve to identify priorities for action in achieving better air quality in Asia.

Air quality measurement capacity

The main objective of air quality monitoring is to collect data to provide information for scientists, policy makers and planners to make informed decisions on managing and improving the environment. Specific objectives may include:

■   the assessment of population or ecosystem exposure to air pollutants
■   determining compliance with statutory air quality criteria
■   providing a sound basis for the development of air quality management plans and/or
■   informing the public about environmental quality.

More detailed technical objectives include the classification of pollution sources, land use planning, traffic planning/management, the determination of spatial and temporal patterns of exposure, the estimation of risks for public health and the environment, or identifying long-term trends. One particular objective of monitoring of air pollutant concentrations is to test compliance with air quality standards. The results of monitoring together with source apportionment can provide feedback for the continuous process of improving air quality by lowering emissions from mobile and stationary sources.

The rationale behind these objectives is to protect human health and the environment. This means that air pollutant concentration levels are considered at those locations where people live. Local levels in the immediate vicinity of power stations and industrial facilities are therefore not included since they are more representative of emissions and less so of human exposure.

A quantitative assessment of the monitoring capabilities of the cities examined is provided by the measurement capability index. This index assesses the ambient air monitoring taking place in each city and the accuracy, precision and representativeness of data produced. Monitoring capabilities have been considered for six pollutants of concern to human health and environment. These are: NO_2, SO_2, PM, CO, O_3 and lead (Pb). Although there is concern over the health impact of other pollutants such as polycyclic aromatic hydrocarbons (PAHs), volatile organic compounds (VOCs), acidic substances, dioxins and polychlorinated biphenyls (PCBs), these have not been addressed as they are not widely or routinely measured in many Asian countries and are currently considered to be of a lower priority.

As indicated earlier, air quality measurements undertaken in cities are dependent on monitoring objectives. The four most frequently used monitoring objectives considered in the context of the questionnaire survey are:

1 assessment of the potential for acute and chronic health effects resulting from exposure to the six pollutants considered, determined by compliance with national air quality standards and international air quality guidelines
2 measurement of trends in pollutants concentrations
3 measurement of the spatial distribution of the pollutants within the city
4 measurement of kerbside concentrations to indicate the contribution of road vehicles and to estimate maximum exposure to traffic-related pollution.

Indicators have been developed to determine the capabilities of cities to meet the above monitoring objectives for the different pollutants. However, no direct indicator has been used for testing compliance with national standards as such standards differ substantially from country to country and have not been established in all countries for all relevant exposure times of all air pollutants in question. Although this is a deficiency in a city's management capability it is not a deficiency in the ability to measure air quality (WHO/UNEP/MARC, 1996). However, if, for example, long-term standards are not set, there is no scale for interpretation of the monitoring results; this fact makes monitoring results less useful. Indicators associated with air quality standards show the capability to assess air quality acceptability and these are considered within the management capabilities index.

Monitoring which is taking place regularly, at least once in six days, is considered as meeting the air measurement capacity index criteria to measure the potential for chronic effects. This index also includes indicators of data validity and representativeness, which can be determined through the quality assurance and control that are currently in place.

Assessment of health effects of pollution

Monitoring at at least one site in a residential area for one year or more is considered as an indicator of the capacity to assess chronic health effects (this does not mean that chronic health effects due to air pollution are assessed in the city). Provision of daily or hourly mean values for each day of a year or more, at at least one site in a residential area, is considered as an indicator of the capacity to assess acute health effects.

Table 2.3 presents the capacity of 20 Asian cities to assess the potential of chronic and acute health effects of different air pollutants. Only six cities have the capacity to assess the potential for both chronic and acute health effects of the six pollutants considered (Bangkok, Beijing, Hong Kong, New Delhi, Shanghai and Taipei). Kathmandu only has the capacity to assess the acute effects of PM and Jakarta can only assess the chronic effects of NO_2, SO_2, PM and CO. Hanoi and Ho Chi Minh City do not appear to be in a position to assess long-term impacts of NO_2, SO_2, PM and Pb,[1] while Metro Manila at present cannot assess chronic effects of NO_2, CO and Pb. As Colombo is not monitoring O_3, it cannot assess the impacts due to this compound. Only eight cities (Bangkok, Beijing, Hong Kong, Kolkata, Mumbai, New Delhi, Shanghai and Taipei) measure the chronic effects of lead, which may reflect the lower concern given to this pollutant as leaded petrol has been phased out in all these cities with the exception of Surabaya.

Table 2.4 shows the capacity to estimate trends for six pollutants. This is where there is at least one site in a residential area in the city monitoring a pollutant for a minimum of five years and it is possible to provide annual mean values.

Nine cities (Bangkok, Beijing, Busan, Hong Kong, Seoul, Shanghai, Singapore, Taipei and Tokyo) estimate trends in all the pollutants while Dhaka, Kathmandu and Surabaya have not yet the capability to estimate trends for any of the pollutants. Kolkata and Mumbai can determine trends for PM, SO_2 and NO_2 with Metro Manila having

Table 2.3 *Capacity to assess the potential for health effects of pollutants*

City	NO₂ Chronic	NO₂ Acute	SO₂ Chronic	SO₂ Acute	PM Chronic	PM Acute	CO Chronic	CO Acute	Pb* Chronic	O₃ Chronic	O₃ Acute
Bangkok	■	■	■	■	■	■	■	■	■	■	■
Beijing	■	■	■	■	■	■	■	■	■	■	■
Busan	■	■	■	■	■	■	■	■	□	■	■
Colombo	■	■	■	■	■	■	■	■	■	■	□
Dhaka	■	■	■	■	■	■	■	■	■	■	■
Hanoi	■	□	□	■	□	■	■	■	■	■	■
Ho Chi Minh City	□	■	■	■	■	■	■	■	■	■	■
Hong Kong	■	■	■	■	■	■	■	■	■	■	■
Jakarta	■	□	■	□	■	□	■	■	■	■	■
Kathmandu	■	□	■	□	■	□	■	■	■	■	■
Kolkata	■	■	■	■	■	■	■	■	□	■	■
Metro Manila	□	■	■	■	■	■	□	■	■	■	■
Mumbai	■	■	■	■	■	■	□	■	■	■	□
New Delhi	■	■	■	■	■	■	■	■	■	■	■
Seoul	■	■	■	■	■	■	■	■	□	■	■
Shanghai	■	■	■	■	■	■	■	■	■	■	■
Singapore	■	■	■	■	■	■	■	■	■	■	■
Surabaya	■	■	■	■	■	■	■	■	□	■	■
Taipei	■	■	■	■	■	■	■	■	■	■	■
Tokyo	■	■	■	■	■	■	■	■	□	■	■

Notes:
* There is no indicator for acute effects of lead since short-term exposure to airborne lead is of little effect.
Chronic: At least one site in a residential area which has been monitoring for one year or more with a frequency of more than one day in six.
Acute: At least one site in a residential area which has been monitoring for one year or more and provides daily or hourly mean values, each day.

■	Compound measured and standard is set
□	Compound not measured or standard not set

Table 2.4 *Capacity to measure trends in pollutant concentrations*

City	PM	O₃*	CO	SO₂	NO₂	Pb
Bangkok	■	■	■	■	■	■
Beijing	■	■	■	■	■	■
Busan	■	■	■	■	■	■
Colombo	■		■	■	■	
Dhaka						
Hanoi	■		■	■	■	■
Ho Chi Minh City	■	■	■	■	■	
Hong Kong	■	■	■	■	■	■
Jakarta	■	■	■	■	■	
Kathmandu						
Kolkata	■			■	■	
Metro Manila	■		■	■	■	
Mumbai	■		■	■	■	
New Delhi	■		■	■	■	■
Seoul	■	■	■	■	■	■
Shanghai	■	■	■	■	■	■
Singapore	■	■	■	■	■	■
Surabaya						
Taipei	■	■	■	■	■	■
Tokyo	■	■	■	■	■	■

Notes: * Annual mean ozone, calculated from all measurements in a year, is not a useful indicator and maximum, an annual mean calculated from eight-hour daylight means, the 98th percentile, second highest value or equivalent statistic should be used.
At least one site in a residential area which has been monitoring for a minimum of five years capable of providing annual mean values.

■	Compound measured
	Compound not measured

time-series for only PM and SO_2. A total of 11 cities are able to determine O_3 trends. However, Colombo, Dhaka, Hanoi, Kathmandu, Kolkata, Metro Manila, Mumbai and New Delhi and Surabaya do not have the capability to estimate O_3 trends.

With regard to the spatial distribution of pollutants, only 13 cities (Bangkok, Beijing, Busan, Ho Chi Minh City, Hong Kong, Jakarta, Metro Manila, New Delhi, Seoul, Shanghai, Singapore, Taipei and Tokyo) have the capacity to measure the spatial distribution of O_3 and seven cities that of Pb (Bangkok, Beijing, Hong Kong, Kolkata, Mumbai, Shanghai and Taipei). Three cities do not determine the spatial distribution of any of the pollutants (Colombo, Dhaka and Surabaya) (see Table 2.5).

Table 2.6 shows that 11 cities (Bangkok, Beijing, Ho Chi Minh City, Hong Kong, Jakarta, Kolkata, Mumbai, New Delhi, Shanghai, Singapore and Taipei) undertake kerbside measurements for five pollutants (PM, CO, SO_2, NO_2 and Pb). Dhaka, Hanoi and Surabaya do not undertake kerbside measurements for any of the pollutants while Kathmandu only measures PM and NO_2.

Quality assurance and quality control (QA/QC) measures are the backbone of any air quality monitoring programme. Therefore, the establishment and implementation of QA/QC programmes and adoption of QA/QC plans ensure that air quality monitoring data (and emissions data and health and environmental monitoring data) are reliable and provide a sound basis for policy making. The quality of the data from monitoring is important to determine whether air quality objectives are being met. Data quality has been assessed here based on the quality assurance and control procedures adopted and reported to be followed by participating cities. Table 2.7 shows that all cities except Hanoi and Jakarta calibrated instruments at least monthly. Flow checks with the certified solutions or gases were performed in all cities except Hanoi, Jakarta, Kolkata, Mumbai and Surabaya. However, such checks are not a guarantee of producing data of known quality. Therefore, auditing of procedures by an independent body is a further requirement for ensuring data quality. Twelve cities have such auditing performed but Colombo, Dhaka, Hanoi, Jakarta, Kathmandu, Kolkata, Mumbai and Surabaya do not. Twelve cities have a sample analysis accredited by a laboratory (Beijing, Busan, Ho Chi Minh City, Hong Kong, Kathmandu, Kolkata, New Delhi, Seoul, Shanghai, Singapore, Taipei and Tokyo). All cities except Kolkata, Mumbai and New Delhi report validation of data once it has been generated although the procedures and accuracy may vary considerably between cities and networks. Fourteen cities (excluding Colombo, Jakarta, Kathmandu, Metro Manila, Surabaya and Taipei) check the location of each site at least every five years to ensure that they are acceptable and meet the objectives of the network.

Table 2.5 *Indicator of the capacity to measure the spatial distribution of pollutants*

City	PM	O₃*	CO	SO₂	NO₂	Pb
Bangkok	■	■	■	■	■	■
Beijing	■	■	■	■	■	■
Busan	■	■	■	■	■	
Colombo						
Dhaka						
Hanoi	■		■	■	■	■
Ho Chi Minh City	■	■	■	■	■	■
Hong Kong	■	■	■	■	■	■
Jakarta	■	■	■	■	■	
Kathmandu	■		■	■	■	■
Kolkata	■		■	■	■	■
Metro Manila	■	■	■	■	■	■
Mumbai	■		■	■	■	■
New Delhi	■	■	■	■	■	■
Seoul	■	■	■	■	■	
Shanghai	■	■	■	■	■	■
Singapore	■	■	■	■	■	■
Surabaya						
Taipei	■	■	■	■	■	■
Tokyo	■	■	■	■	■	

Notes: At least three sites, one site in each of a predominantly residential, commercial and industrial area of the city, which have been monitoring for at least one year using equivalent equipment and methodologies (or those for which inter-comparisons have been conducted), with a monitoring frequency greater than one day in six, for the above pollutants.

* The O₃ sites should be located upwind and downwind in the suburbs of the city, and in the city centre, due to the secondary nature of O₃ pollution.

If mapping of pollutants had been conducted using modelling and an emissions inventory, this would be considered as meeting the indicator's criteria.

■ Compound measured

☐ Compound not measured

Table 2.6 *Indicator of the capacity to measure kerbside concentrations.*

City	PM	CO	SO$_2$	NO$_2$	Pb
Bangkok	■	■	■	■	■
Beijing	■	■	■	■	■
Busan	■	■	■	■	□
Colombo	■	■	■	■	□
Dhaka	□	□	□	□	□
Hanoi	□	□	□	□	□
Ho Chi Minh City	■	■	■	■	■
Hong Kong	■	■	■	■	■
Jakarta	■	■	■	■	■
Kathmandu	■	□	■	■	■
Kolkata	■	■	■	■	■
Metro Manila	■	■	■	□	■
Mumbai	■	■	■	■	■
New Delhi	■	■	■	■	■
Seoul	■	■	■	■	□
Shanghai	■	■	■	■	■
Singapore	■	■	■	■	■
Surabaya	□	□	□	□	□
Taipei	■	■	■	■	■
Tokyo	■	■	■	■	□

Notes: There is no indicator for O$_3$ since concentrations normally are very low at the roadside due to depletion by reaction with NO.
A site monitoring within 3 m of the roadside or kerb operating for one year or more at least one day in six, for the above pollutants.

■	Compound measured
□	Compound not measured

Table 2.7 *Quality assurance and control procedures followed*

City	Monthly calibration	Flow checks with certified gases	Auditing by an independent body	Sample analysis by accredited laboratory	Data validation	Site reviews at least every 5 years
Bangkok	■	■	■	■	■	■
Beijing	■	■	■	■	■	■
Busan	■	■	■	■	■	■
Colombo	■	■	□	□	■	■
Dhaka	■	■	□	□	■	■
Hanoi	□	□	■	■	■	■
Ho Chi Minh City	□	□	■	■	■	■
Hong Kong	■	■	■	■	■	■
Jakarta	■	■	■	□	■	■
Kathmandu	■	■	□	■	■	■
Kolkata	■	■	■	■	□	■
Metro Manila	■	■	■	■	□	□
Mumbai	■	■	■	■	■	■
New Delhi	■	■	■	■	□	■
Seoul	■	■	■	■	■	■
Shanghai	■	■	■	■	■	■
Singapore	■	■	■	■	■	■
Surabaya	■	■	■	■	■	□
Taipei	■	■	■	■	■	□
Tokyo	■	■	■	■	■	■

■ Procedure conducted

□ Procedure not conducted

Figure 2.3 shows the scores for the air quality measurement capability index for each city. Eleven cities (Bangkok, Beijing, Busan, Ho Chi Minh City, Hong Kong, New Delhi, Seoul, Shanghai, Singapore, Taipei and Tokyo) are classified by the index as having excellent measurement capabilities; six are moderate (Colombo, Hanoi, Jakarta, Kolkata, Metro Manila and Mumbai), two limited (Dhaka and Surabaya) and one minimal (Kathmandu). Those cities with excellent status are those which have an established AQM network capability of monitoring six key pollutants.

Data assessment and availability

In addition to QA/QC it is also important that the information, once validated and assessed, are widely disseminated in an appropriate format for decision makers who are responsible for maintaining acceptable air quality levels. Widespread access to data and information enables inputs to be made to the decision-making process and encourages transparent decision making (WHO/UNEP/MARC, 1996).

Capability in data assessment

The index of data assessment and availability assesses the statistical operations and data assessment performed on raw data and public access to this data and information.

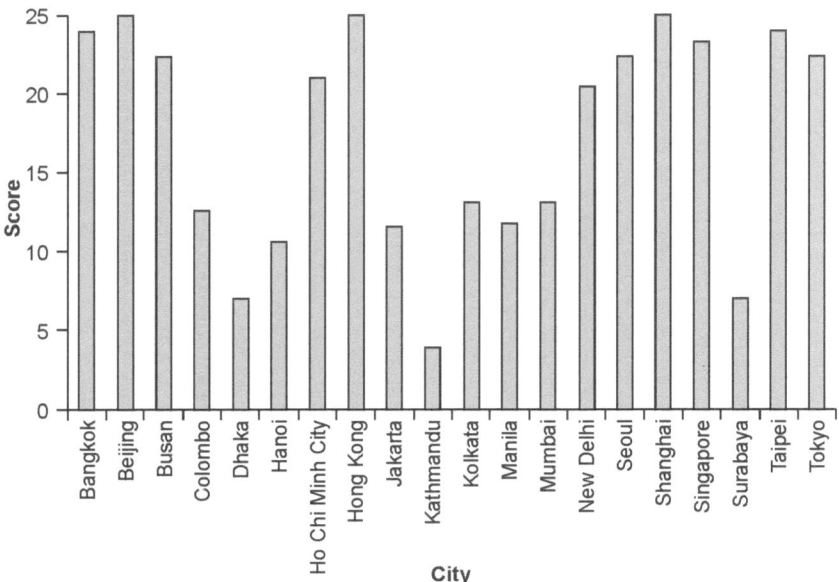

Figure 2.3 *Air quality measurement index*

This index is a measurement of the data analysis performed by cities rather than those assessments for which data are available and could be conducted – this is assessed by the measurement index.

Indicators of data assessment reflect the data analysis performed by the city and how this analysis is achieved. Data availability indicators, on the other hand, determine where information is published, whether and how it is reported to the public and if warnings and advice are issued to the public during pollution episodes.

Table 2.8 shows the assessments which are undertaken on air quality data in each of the cities. It demonstrates that simple data analysis procedures, such as mean and maximum values, are performed by all cities participating in the study. However, only five cities (Bangkok, Hong Kong, Shanghai, Singapore and Tokyo) undertook percentiles. All cities except Dhaka, Kathmandu and Surabaya undertake trend analyses in pollutant concentrations. An analysis of compliance with national air quality standards, USEPA standards or WHO guidelines is undertaken by all cities. Nine cities (Beijing, Busan, Ho Chi Minh City, Hong Kong, Seoul, Shanghai, Singapore, Taipei and Tokyo) undertake mapping of the spatial distribution of pollutants. Eleven cities undertake prediction modelling (Bangkok, Beijing, Hanoi, Ho Chi Minh City, Hong Kong, Metro Manila, Shanghai, Singapore, Surabaya, Taipei and Tokyo). Exposure assessment, the essential ingredient for epidemiological studies, is performed by only seven cities (Bangkok, Beijing, Hong Kong, Shanghai, Singapore, Taipei and Tokyo) while 11 cities (Bangkok, Beijing, Hong Kong, Metro Manila, Mumbai, New Delhi, Seoul, Shanghai, Singapore, Taipei and Tokyo) report epidemiological studies being performed.

Data dissemination

Table 2.9 shows the mechanism by which air quality data is disseminated in the cities. All cities except Jakarta, Mumbai, New Delhi and Shanghai have raw air quality data available. Eleven cities (Bangkok, Ho Chi Minh City, Hong Kong, Kathmandu, Kolkata, New Delhi, Seoul, Shanghai, Singapore, Taipei and Tokyo) formally publish data in reports. All cities except Busan, Dhaka, Hanoi, Ho Chi Minh City and Metro Manila use the media, either press, television or radio, to release information on pollution levels to the public. The issuing of air quality alerts is important for vulnerable groups. Acute exposure to SO_2, PM, NO_2, CO and O_3 can affect individuals with respiratory or cardiovascular disease. The issuing of advice to such groups can directly reduce the consequences of a pollution episode on human health without affecting the actual levels of pollution (WHO/UNEP/MARC, 1996). However, only ten cities (Beijing, Busan, Hong Kong, Jakarta, Kolkata, Seoul, Shanghai, Singapore, Taipei and Tokyo) issue air quality alerts during periods of poor air quality. Thirteen

Table 2.8 *Air quality assessments*

City	Simple statistics	Percentiles	Compliance	Trends	Mapping	Prediction modelling	Exposure assessment	Epidemiological studies
Bangkok	✓	✓	✓	✓		✓	✓	✓
Beijing	✓		✓	✓		✓	✓	✓
Busan	✓		✓	✓	✓			
Colombo	✓		✓	✓				
Dhaka	✓							
Hanoi	✓	✓	✓	✓	✓	✓		
Ho Chi Minh City	✓		✓					
Hong Kong	✓	✓	✓	✓	✓	✓		✓
Jakarta	✓		✓	✓				✓
Kathmandu	✓		✓	✓				
Kolkata	✓		✓	✓				
Metro Manila	✓		✓	✓	✓	✓		✓
Mumbai	✓		✓	✓				✓
New Delhi	✓		✓	✓				✓
Seoul	✓		✓	✓	✓			
Shanghai	✓	✓	✓	✓			✓	✓
Singapore	✓	✓	✓					
Surabaya	✓		✓	✓		✓		
Taipei	✓							✓
Tokyo								✓

Assessment conducted
Assessment not conducted

Table 2.9 Data dissemination

City	Raw data	Published reports	Reported in the media	Air quality alerts issued	Information boards in city centre
Bangkok	■	■	■		■
Beijing	■		■	■	
Busan	■			■	■
Colombo	■		■		■
Dhaka	■				
Hanoi					
Ho Chi Minh City	■	■			■
Hong Kong	■		■	■	■
Jakarta			■	■	■
Kathmandu	■		■		■
Kolkata	■		■		■
Metro Manila	■		■		
Mumbai				■	■
New Delhi	■		■	■	■
Seoul	■		■	■	■
Shanghai				■	
Singapore	■				
Surabaya	■		■		■
Taipei				■	■
Tokyo					■

■ Air quality information available

☐ Air quality information not available

cities provide air quality information on information boards in city centres. Beijing, Dhaka, Hanoi, Hong Kong, Metro Manila, Shanghai and Singapore do not.

Figure 2.4 shows the overall capacity of the cities to make sufficient use of their air quality data. Seven cities are considered excellent (Bangkok, Beijing, Hong Kong, Shanghai, Singapore, Taipei and Tokyo); one is good (Seoul); six are moderate (Busan, Ho Chi Minh City, Jakarta, Kolkata, Mumbai and New Delhi); five limited (Dhaka, Hanoi, Kathmandu, Manila and Surabaya) and one minimal (Colombo). The scores for this index are generally lower than those for the capability to measure ambient air pollution, which indicates that cities could make better use of their data.

Emission estimates

Quantitative and reliable knowledge of the sources of various pollutant emissions is an important component of AQM. Emission estimates through the compilation of emission inventories enable better targeting of control measures for major sources of emissions. Without emission inventories, regulatory action may be taken on sources which are not the most polluting and the emission reduction is not achieved. Emission estimates allow the spatial distribution of emission across the city to be determined and with the use of dispersion modelling pollution hotspots or areas of poor air quality can be identified.

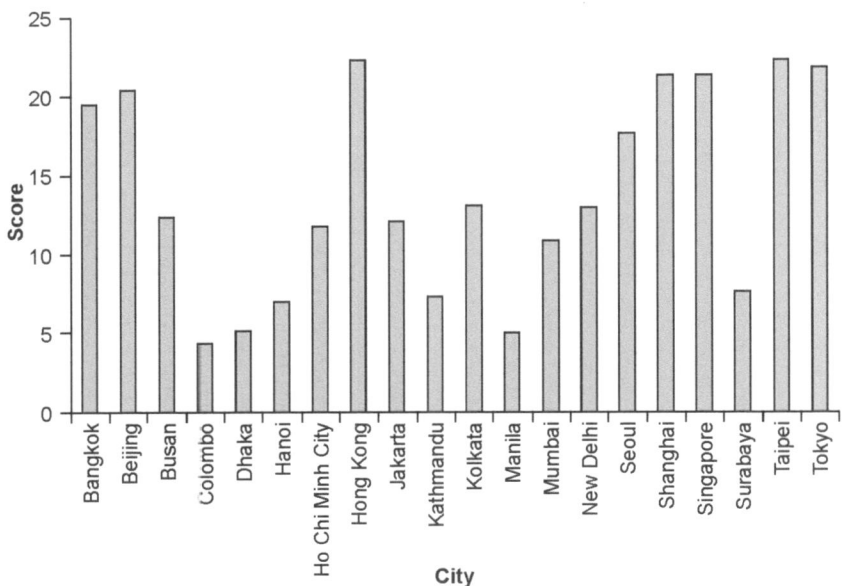

Figure 2.4 *Air quality assessment and availability index scores*

Capability in emission estimates

Table 2.10 shows the major source categories each city includes in its emissions inventory. The majority of the cities have made some estimates of industrial emissions and car emissions (excluding Dhaka for industrial and Surabaya for mobile sources). Most cities have estimated emissions from motorcycles and heavy goods vehicles and buses (excluding Beijing, Hanoi, Ho Chi Minh City, Kathmandu and Surabaya) and domestic/commercial sources (excluding Dhaka, Kolkata, Mumbai and New Delhi). With regard to emission estimates from energy sources, five cities have not undertaken any estimates (Dhaka, Hanoi, Ho Chi Minh City, Kathmandu and Taipei). Only seven cities estimated emissions from other sources such as aircraft and ships (Bangkok, Busan, Hong Kong, Seoul, Singapore, Taipei and Tokyo).

Table 2.11 shows that all cities except Hanoi, Kathmandu, Kolkata, Mumbai and Surabaya have undertaken emission estimates for at least five of the pollutants. Only four cities undertake emission estimates for Pb (Busan, Dhaka, Hong Kong and Taipei). The reason for this observation may be that most cities do not consider Pb a pollutant of concern since lead was phased out in gasoline. Five cities did not provide estimates for hydrocarbons (HC) (Hanoi, Kathmandu, Kolkata, Mumbai and Surabaya).

Information on emissions needs to be decision relevant and therefore requires a minimum level of certainty. However, only eight cities (Bangkok, Busan, Hong Kong, Seoul, Singapore, Surabaya, Taipei and Tokyo) have estimates of emissions based on actual measurements (see Table 2.12), which suggests that the degree of uncertainty in the emission estimates in 12 cities may be substantial. Twelve cities (Bangkok, Beijing, Busan, Hanoi, Hong Kong, Jakarta, Metro Manila, New Delhi, Seoul, Singapore, Taipei, Tokyo) include estimates of non-combustion sources (e.g. waste deposits).

Validation of emissions inventories involves ascertaining the completeness of data for each sector and activity and consistency of emissions factors and estimates relative to those applied in other countries of similar size and level of industrialization and development. Validation of emissions inventories is essential to produce data of known quality and to ensure that it is adequate for its intended use. Only six cities (Beijing, Busan, Seoul, Singapore, Taipei and Tokyo) validate their emission estimates while two cities (Bangkok and Metro Manila) actually publish the emission data in full. Non-validated emission inventories produce data of unknown quality and value and are equivalent to monitoring data with no quality control and assurance (WHO/UNEP/MARC, 1996). Emission inventories have to be regularly reviewed and updated in order to ensure the efficiency of measures taken and decisions made. Only half of the cities (Beijing, Busan, Ho Chi Minh City, Hong Kong, Metro Manila, Seoul, Shanghai, Singapore, Taipei and Tokyo) review their emission inventories every two years.

Table 2.10 *Emission estimates for major source categories*

City	Industrial	Energy	Cars	Motor cycles	HGV/ Buses	Other mobile sources	Domestic/ commercial
Bangkok	■	■	■	■	■	■	■
Beijing	■	■	■	□	■	■	■
Busan	■	■	■	■	■	■	■
Colombo	■	■	■	■	■	■	■
Dhaka	□	□	■	■	■	□	□
Hanoi	■	□	■	■	■	■	■
Ho Chi Minh City	■	□	■	■	■	■	■
Hong Kong	■	■	■	■	■	■	■
Jakarta	■	■	■	■	■	□	■
Kathmandu	■	■	■	□	■	■	■
Kolkata	■	■	■	■	■	■	■
Metro Manila	■	■	■	■	■	■	■
Mumbai	■	■	■	■	■	■	□
New Delhi	■	■	■	■	■	■	□
Seoul	■	■	■	■	■	■	■
Shanghai	■	■	■	■	■	□	■
Singapore	■	■	■	■	■	■	■
Surabaya	■	■	□	■	■	■	■
Taipei	■	□	■	■	■	■	■
Tokyo	■	■	■	■	■	■	■

■ Emission estimate available
□ Emission estimate not available

Table 2.11 *Pollutants for which emission estimates have been derived*

City	PM	CO	SO$_2$	NO$_x$	HC	Pb
Bangkok	✓	✓	✓	✓	✓	
Beijing	✓	✓	✓	✓	✓	
Busan	✓	✓	✓	✓	✓	✓
Colombo	✓	✓	✓	✓	✓	
Dhaka	✓	✓	✓	✓	✓	✓
Hanoi	✓	✓	✓	✓		
Ho Chi Minh City	✓	✓	✓	✓	✓	
Hong Kong	✓	✓	✓	✓	✓	
Jakarta	✓	✓	✓	✓	✓	
Kathmandu	✓					
Kolkata	✓		✓		✓	
Metro Manila	✓	✓	✓	✓	✓	
Mumbai			✓	✓	✓	✓
New Delhi	✓	✓	✓	✓	✓	
Seoul	✓	✓	✓	✓	✓	
Shanghai	✓	✓	✓	✓	✓	
Singapore	✓	✓	✓	✓	✓	
Surabaya	✓	✓	✓	✓		
Taipei	✓	✓	✓	✓	✓	✓
Tokyo	✓	✓	✓	✓	✓	

✓ = Emission estimate derived (shaded)

(blank) = Emission estimate not derived

Table 2.12 *Emission estimates validity and availability*

City	Non-combustion sources included	Based on actual measurement	Validated	Two-year review	Published in full
Bangkok	●	●			●
Beijing	●		●	●	
Busan	●	●	●	●	
Colombo					
Dhaka					
Hanoi	●				
Ho Chi Minh City				●	
Hong Kong				●	
Jakarta	●				
Kathmandu					
Kolkata					
Metro Manila				●	●
Mumbai					
New Delhi					
Seoul	●	●	●	●	
Shanghai				●	
Singapore	●		●	●	
Surabaya		●			
Taipei	●	●	●	●	
Tokyo	●	●	●	●	

● Procedure conducted

☐ Procedure not conducted

Figure 2.5 illustrates the diverse capacity among the cities to estimate emissions. Using the index, six cities have an excellent capability (Busan, Hong Kong, Seoul, Singapore, Taipei and Tokyo), six are rated as good (Bangkok, Colombo, Jakarta, Metro Manila, New Delhi and Shanghai), four as moderate (Beijing, Kolkata and Mumbai) and four as limited (Hanoi, Ho Chi Minh City, Kathmandu and Surabaya). At least eight cities with a score below 15 have a significant deficiency in making the information useful for the development of rational AQM strategies.

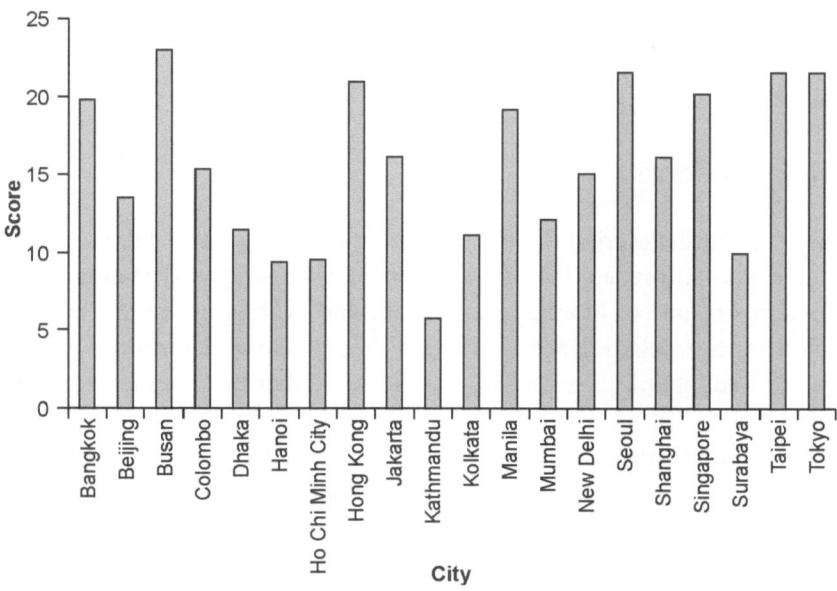

Figure 2.5 *Emission estimation capability*

MANAGEMENT ENABLING CAPABILITIES

AQM aims to maintain the quality of the air that protects human health and well-being but also provides protection of animals, plants (crops, forests and natural vegetation), ecosystems, materials and aesthetics, such as natural levels of visibility. AQM is a tool which enables government authorities to set objectives to achieve and maintain clean air and reduce the impacts of air pollution on human health and the environment.

AQM enables government authorities, in collaboration with other stakeholders, to identify and establish appropriate policies and regulatory frameworks for air quality. It enables all major sources of air pollution to be identified and objectives and targets for human and environmental health to be set. AQM establishes the institutional structure and programmes to implement policies and achieve objectives and targets such as monitoring and urban planning.

The management-enabling capacity indicators are intended to assess the basic AQM tools which are used to control air pollution in each Asian city. These indicators are associated with how the first three capabilities (AQ measurement, AQ data assessment and availability, and emission estimates) are applied to develop air quality control strategies. These indicators do not relate to the relative success of specific strategies introduced. Rather, they refer to the existence and application of air quality standards and guidelines, the presence of emission controls and whether or not adequate capabilities exist to ensure their enforcement. The aim is to determine whether air quality information is used in planning and designing AQM strategies and programmes in the city (WHO/UNEP/MARC, 1996).

Air quality standards

Air quality standards describe the acceptable level of air quality adopted by a regulatory authority as enforceable. In its simplest form, an air quality standard should be defined in terms of one or more concentrations and averaging times. The number of times a short-term standard may be exceeded should also be determined. Once standards have been adopted, it is necessary to assess whether the level is being exceeded or complied with.

Table 2.13 shows that, with few exceptions, all cities use national ambient air quality standards or international guidelines for all air pollutants of concern to assess the acceptability of air quality. Tokyo does not have a long-term standard for Pb, possibly due to the early phasing out of leaded gasoline. The Indian cities, Kathmandu and Metro Manila do not have a standard for O_3. Colombo, Hanoi and Ho Chi Minh City[1] do not have long-term standards for SO_2 and NO_2. All cities have regulations to enforce their air quality standards or take action in order to ensure that exceedances do not routinely occur.

Capacity to impose emission controls and penalties and use of air quality information

An AQM system requires the introduction of emission limits on stationary and mobile sources and the monitoring of emissions to ensure they comply with such limits. Emission monitoring is therefore an important component of management capability. Moreover, information on local air quality should be considered in the planning and development of new roads and industries. Table 2.14 shows that the majority of the cities (excluding Colombo, Jakarta and Kathmandu) have emission limits and controls

Table 2.13 *Air quality standards*

City	PM		O₃	CO	SO₂		NO₂		Pb	Regulations to enforce compliance
	Acute	Chronic	Acute	Acute	Acute	Chronic	Acute	Chronic	Chronic	
Bangkok	✓	✓	✓	✓	✓	✓	✓	✓	✓	✓
Beijing	✓	✓	✓	✓	✓	✓	✓	✓	✓	✓
Busan	✓	✓	✓	✓	✓	✓	✓	✓	✓	✓
Colombo	✓	✓	✓	✓	✓		✓		✓	✓
Dhaka	✓	✓	✓	✓	✓	✓	✓	✓	✓	✓
Hanoi[1]	✓		✓	✓	✓		✓	✓	✓	✓
Ho Chi Minh City[1]		✓	✓	✓	✓		✓	✓	✓	✓
Hong Kong	✓	✓	✓	✓	✓	✓	✓	✓	✓	✓
Jakarta	✓	✓	✓	✓	✓	✓	✓	✓	✓	✓
Kathmandu	✓	✓		✓	✓	✓	✓	✓	✓	✓
Kolkata	✓	✓		✓	✓	✓	✓	✓	✓	✓
Metro Manila	✓	✓		✓	✓	✓	✓	✓	✓	✓
Mumbai	✓	✓		✓	✓	✓	✓	✓	✓	✓
New Delhi	✓	✓		✓	✓	✓	✓	✓	✓	✓
Seoul	✓	✓	✓	✓	✓	✓	✓	✓	✓	✓
Shanghai	✓	✓	✓	✓	✓	✓	✓	✓	✓	✓
Singapore	✓	✓	✓	✓	✓	✓	✓	✓	✓	✓
Surabaya	✓	✓	✓	✓	✓	✓	✓	✓	✓	✓
Taipei	✓	✓	✓	✓	✓	✓	✓	✓	✓	✓
Tokyo	✓	✓	✓	✓	✓	✓	✓	✓		✓

✓ Air quality standard set

    No air quality standard

set for key pollution sources. Additional emission controls in case of episodes with extremely high air pollutant concentrations are imposed in only three cities (Hong Kong, Singapore and Tokyo). Thirteen cities actually impose penalties for exceedance of emission limits for both industrial and mobile sources while Jakarta, Kathmandu and Mumbai impose limits for only mobile sources. There is no certainty, however, as to what extent penalties are actually enforced.

Only four cities (Hong Kong, Singapore, Taipei and Tokyo) take into consideration air quality information in the planning of new roads. Seven cities (Colombo, Dhaka, Jakarta, Kathmandu, Metro Manila, New Delhi and Surabaya) do not consider air quality information in the planning of new industries.

Figure 2.6 provides an overview of the capability of cities to utilize AQM tools such as ambient air quality standards and emissions regulations. A total of six cities (Busan, Hong Kong, Seoul, Singapore, Taipei and Tokyo) have excellent ratings; seven good (Bangkok, Beijing, Kolkata, Metro Manila, Mumbai, New Delhi and Shanghai), three moderate (Ho Chi Minh City, Jakarta and Surabaya) and four limited (Colombo, Dhaka, Hanoi and Kathmandu). Many of the cities in the study have introduced some of the management tools required to formulate and implement AQM strategies.

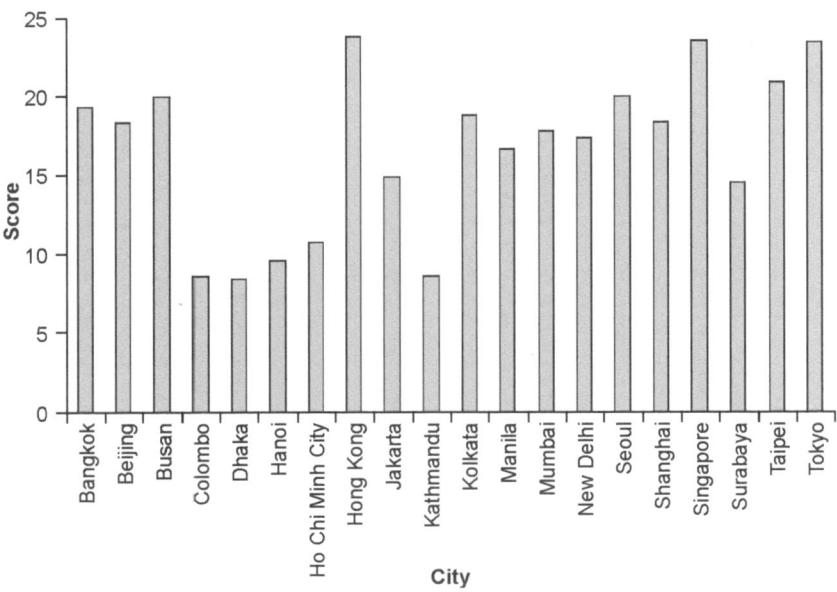

Figure 2.6 *Air quality management tools index*

Table 2.14 *Emission controls, penalties imposed for no compliance, land use planning*

	Emission controls		Additional emission controls during episodes	Penalties imposed		Consideration of local air quality in planning	
	Industrial	Mobile		Industrial	Mobile	Roads	Industrial plants
Bangkok	■	■		■	■		■
Beijing	■	■		■	■		■
Busan	■	■		■	■		■
Colombo		■		■	■		■
Dhaka	■	■		■	■		
Hanoi	■	■		■	■		■
Ho Chi Minh City	■	■		■	■		■
Hong Kong	■	■	■	■	■	■	■
Jakarta		■		■	■	■	■
Kathmandu		■		■	■		■
Kolkata	■	■		■	■		■
Metro Manila	■	■		■	■	■	■
Mumbai	■	■		■	■		
New Delhi	■	■		■	■	■	
Seoul	■	■		■	■		■
Shanghai	■	■		■	■	■	■
Singapore	■	■	■	■	■	■	■
Surabaya	■	■		■	■		
Taipei	■	■		■	■	■	■
Tokyo	■	■	■	■	■	■	■

■ Procedure conducted
☐ Procedure not conducted

OVERALL MANAGEMENT CAPABILITY

This chapter has brought together the information supplied by the 20 Asian cities to provide an overview of AQM capability and to identify where poor capability exists both generally and specifically. An overview of the capacity of cities to formulate and implement AQM strategies is provided by combining the scores from the four component indices (see Figure 2.7). A wide range of scores and capabilities exist within the 20 Asian cities. Seven cities achieved an excellent rating (Bangkok, Hong Kong, Seoul, Singapore, Shanghai, Taipei and Tokyo). Beijing, Busan and New Delhi were rated with a good overall capability. A total of six cities were rated with an overall moderate capacity in AQM (Colombo, Ho Chi Minh City, Jakarta, Kolkata, Metro Manila and Mumbai). Dhaka, Hanoi, Kathmandu and Surabaya have limited AQM capability. None of the cities considered had a minimal capability.

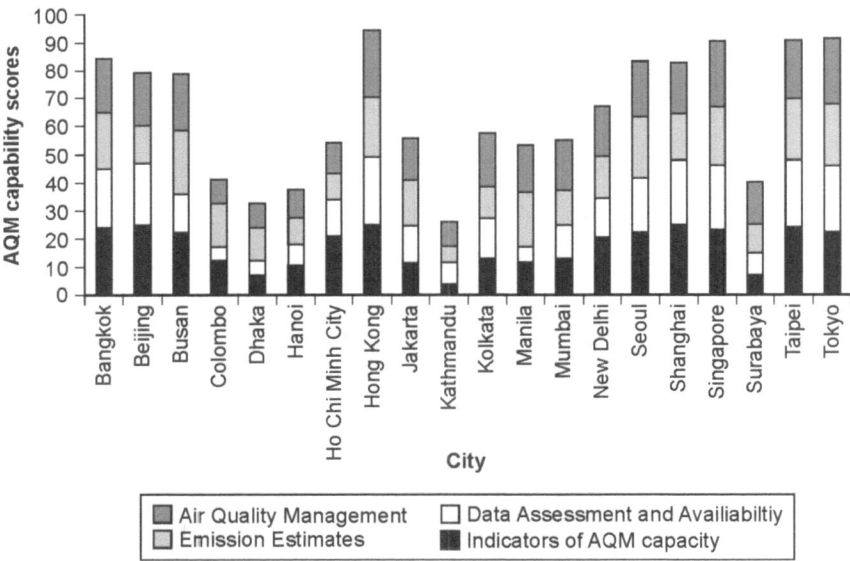

Figure 2.7 *Overall air quality management index tools*

Only Hong Kong, Singapore, Taipei and Tokyo have an excellent rating in every element of AQM capability. All cities have some level of useful capability; however, at least five cities require further capability for effective AQM. The overall performance is summarized in Table 2.15.

Table 2.15 *AQM capability in 20 Asian cities*

Overall capability	City
Excellent	Bangkok, Hong Kong, Seoul, Singapore, Shanghai, Taipei, Tokyo
Good	Beijing, Busan, New Delhi
Moderate	Colombo, Ho Chi Minh City, Jakarta, Kolkata, Metro Manila, Mumbai
Limited	Dhaka, Hanoi, Kathmandu, Surabaya
Minimal	None

CONCLUSION

Participation in the assessment required city collaboration to complete a detailed questionnaire survey and review a city profile. The results from the questionnaire survey were cross-checked with existing studies and data and, where necessary, additional information and clarification was sought. The survey was undertaken in the period 2002–2004 and additional clarification and updates were made until May 2005. While every effort was made to clarify the information obtained from the questionnaire survey, the collaborative methodology used means that the cities may have achieved higher scoring of their capability for AQM than exists in reality. Similarly, while every effort has been made in updating the data changes, improvements in AQM within the cities may have been made since this assessment and are therefore not considered here. Future studies would need to address further the independent assessment of the information obtained from the city authorities. The questionnaire survey used to assess future AQM capability will also need to be adapted to include new priorities in AQM.

Despite the potential limitations of such an approach, this chapter has demonstrated the key components of a comprehensive AQM system and the range of AQM capability that exists in a selected number of Asian cities of varying size and level of economic development. In the course of time and development, AQM capability, and consequently scores, may increase due to the installation of more extensive air quality monitoring networks and application of the more sophisticated tools of AQM. An example is the fully automated Metro Manila air quality monitoring network that has been in operation since 2004. This network system has replaced the non-coordinated monitoring activities of the Environmental Management Bureau, Department of Environment and Natural Resources, as well as those of the Manila Observatory and

	Measurement of air quality	Data assessment and availability	Emission inventory	Air quality management	Overall score
Bangkok					
Beijing					
Busan					
Colombo					
Dhaka					
Hanoi					
Ho Chi Minh City					
Hong Kong					
Jakarta					
Kathmandu					
Kolkata					
Metro Manila					
Mumbai					
New Delhi					
Seoul					
Shanghai					
Singapore					
Surabaya					
Taipei					
Tokyo					

Minimal   Limited   Moderate   Good   Excellent

Figure 2.8 *Summary of AQM capabilities in 20 Asian cities*

the Philippine Nuclear Research Institute in order to fully maximize the benefits from these different monitoring initiatives. Figure 2.8 provides a summary of each city with respect to the four component elements of management capability.

NOTES

1 This refers to the standards of TCVN 5937-1995 which defines only 1-hour and 24-hour standards for the compounds in question. A more recent regulation, TCVN 5937-2005 also sets annual standards, which, however, have not yet been enforced.

REFERENCES

BEST (2000) 'The State of the Art of Benchmarking', BEST – Benchmarking European Sustainable Transport, 19–20 October 2000, Brussels, www.bestransport.org/cadreconf7.html accessed in February 2006

BEST (2003) 'Final BEST Conference – Conference 6 Report', 12–13 March 2003, Brussels, www.bestransport.org/cadreconf7.html accessed in February 2006

CAI-Asia (2006) 'Clean Air Initiative for Asian Cities', www.cleanairnet.org/caiasia/1412/channel.html accessed in April 2006

EEA (1998) *Assessment and Management of Urban Air Quality in Europe*, EEA Monograph No. 5, Copenhagen, Denmark, European Environment Agency

EU (1999) 'Council Directive 1999/30/EC of 22 April 1999 relating to Limit Values for Sulphur Dioxide, Nitrogen Dioxide and Oxides of Nitrogen, Particulate Matter and Lead in Ambient Air', *Official Journal*, L163, 29/06/1999, pp0041–0060

EU (2000) 'Directive 2000/69/EC of the European Parliament and of the Council of 16 November 2000 relating to Limit Values for Benzene and Carbon Monoxide in Ambient Air', *Official Journal*, L 313, 13/12/2000, pp0012–0021, http://europa.eu.int/eur-lex/lex/LexUriServ/LexUriServ.do?uri=CELEX:32000L0069:EN:HTML accessed in April 2006

EU (2002) 'Council Directive 2002/3/EC of the European Parliament and of the Council of 12 February 2002 relating to Ozone in Ambient Air', *Official Journal*, L067, 09/03/2002, pp0014–0030

Gudmundsson, H. (2003) 'Benchmarking and European sustainable transport Policies', *World Transport Policy & Practice*, vol 9, no 2, pp24–31, www.ecologica.co.uk/WTPParticles.html#Volume9 accessed in February 2006

OECD (2002) *Benchmarking Intermodal Freight Transport*, Paris, Organisation for Economic Co-operation and Development

The Times (2003) *Comprehensive Time Atlas*, 11th edn, London, Time Books

UN (2004) 'World Urbanization Prospects: The 2003 Revision', ST/ESA/SER.A/237, United Nations, Department of Economic and Social Affairs, Population Division, New York, www.un.org/esa/population/publications/wup2003 accessed in February 2006

UNEP/WHO (1992) *Urban Air Pollution in Megacities of the World*, United Nations Environment Programme/World Health Organization, Oxford, Blackwell

USEPA (1997) 'National Ambient Air Quality Standards (NAAQS)', United States Environmental Protection Agency, www.epa.gov/air/criteria.html accessed in February 2006

USEPA (2006) 'Review of the Process for Setting National Ambient Air Quality Standards', NAAQS Process Review Workgroup, United States Environmental Protection Agency, Research Triangle Park, NC, www.epa.gov/ttn/naaqs/naaqs_process_report_march2006.pdf accessed in April 2006

WHO (2000) 'Guidelines for Air Quality', WHO/SDE/OEH/00.02, World Health Organization, Geneva, www.who.int/peh/ accessed in February 2006

WHO (2005) 'WHO Air Quality Guidelines Global Update 2005', Report on a Working Group meeting, Bonn, Germany, 18–20 October 2005, Copenhagen, World Health Organization, WHO Regional Office for Europe

WHO/UNEP/MARC (1996) *Air Quality Management and Assessment Capabilities in 20 Major Cities*, UNEP/DEIA/AR.96.2/WHO/EOS.95.7, Nairobi, Kenya, United Nations Environment Programme

three

Air Quality Management in Twenty Asian Cities

INTRODUCTION

This chapter provides a summary of the current state of AQM in 20 Asian cities. It synthesizes for each city the key data and information relevant to AQM and provides general geographical, demographic and meteorological data. The estimated population in 2003 and the projected population in 2015 is taken from UN sources (UN, 2004) while the climate is classified according to the Köppen system (The Times, 2003).

The city summaries outline the most important legislation related to AQM in each city and provide an overview of pollutant emissions and monitoring of key pollutants such as CO, SO_2, TSP, PM_{10}, NO_x, O_3 and lead. Where possible, national air quality standards are compared with WHO guideline values (WHO, 2000; 2005), United States air quality standards (USEPA, 1997) and the EU limit values (EU, 1999, 2000, 2002). For each city the average concentrations of pollutants monitored is given for 2003 or 2004. These were the most recent air quality data for the cities that were available to the authors. Where possible, ambient concentrations are compared with national standards. Finally, the city summaries refer to major epidemiological studies undertaken in the cities highlighting the impact of air pollution on human health.

Table 3.1 *Cities categorized according to climate*

Cooler humid	Cool summer	
	Continental	Beijing
	Warm summer	Seoul
Warmer humid	Temperate	
	Humid	Tokyo
	Sub-tropical	Busan, Hanoi, Hong Kong, Kathmandu, Shanghai, Taipei, Tokyo
	Mediterranean	
Dry	Steppe	New Delhi
	Desert	
	Savannah	Bangkok, Ho Chi Minh City, Kolkata, Mumbai
Tropical humid	Rain forest	Colombo, Dhaka, Jakarta, Metro Manila, Singapore, Surabaya

Source: The Times (2003)

BANGKOK

UN estimated population 2003: 6.49 million
UN projected population 2015: 7.46 million
Area: 1569 km²
Climate: Tropical savannah
Map reference: 13° 44'N, 100° 33'E
Annual mean precipitation: 1500 mm
Altitude: 2.31 m
Annual mean temperature range: 20–35°C

SITUATIONAL ANALYSIS

General

Bangkok is the capital city of Thailand. It is situated on the low flat plane of Chao Phraya River which extends to the Gulf of Thailand. The city covers an area of 1568.737 km² and is 2.31 metres above sea level (UNEP/BMA, 2004). Bangkok has a tropical savannah climate.

Legislation

The 1992 Enhancement and Conservation of National Environmental Quality Act (NEQA) outlines the responsibility of the Pollution Control Department (PCD) in determining pollution control areas and establishing ambient air quality and emission standards. The Act required the establishment of a Pollution Control Committee chaired by the Permanent Secretary of the Ministry of Natural Resources and Environment (MONRE). The Committee is responsible for the formulation of plans, policies and pollution prevention measures (World Bank, 2002; Mottershead, 2002). Thailand has adopted ambient air quality standards and emission standards for both stationary and mobile sources.

National ambient air quality standards exist for all main pollutants. The 24-hour PM_{10} standard of 120 µg/m³ in Bangkok is stricter than the USEPA standard (150 µg/m³) but more lenient than the EU limit value (50 µg/m³). The annual means are the same for Thailand and the US (50 µg/m³), while the EU limit value is slightly lower (40 µg/m³). The Thai SO_2 standard of 314 µg/m³ (120 ppb) for 24-hour averaging time is lenient compared with the EU limit value (125 µg/m³) and WHO guideline (20 µg/m³); the same is true for the annual standard of 105 µg/m³ (40 ppb) as compared to the USEPA standard of 78 µg/m³. The lead standard is not directly comparable to the WHO guidelines and USEPA standards because of differences in the averaging time. The 8-hour CO standard of 10,305 µg/m³ (9000 ppb) is practically the same as the USEPA standard and slightly more lenient than the WHO guideline and EU limit values (10,000 µg/m³). The 1-hour CO standard of Thailand of 34,200 µg/m³ (30,000 ppb) is more lenient than the WHO 1-hour guideline value (30,000 µg/m³) and practically the same as the USEPA standard (35,000 µg/m³).

For mobile sources Euro II equivalent standards for new vehicles were adopted in 2001 and Euro III standards for light-duty vehicles in 2004. Correspondingly, levels of sulphur in fuels have been reduced. Mobile source emission and fuel standards in Thailand are more advanced compared to other developing countries in Asia. For stationary sources, Thailand has introduced desulphurization units in power plants and has set emission standards for key sources (Mottershead, 2002).

Emissions

Mobile sources contribute a substantial amount to the total NO_x (80 per cent), CO (75 per cent) and PM (54 per cent) in the Bangkok Metropolitan Region (BMR) while stationary sources account for approximately 96 per cent of the total SO_2 and area sources for 94 per cent of total VOC emissions (PCD, 2000). With respect to PM_{10}, mobile sources and re-suspended dust from roads contribute approximately 51 per cent, industrial and commercial boilers 34 per cent, power plants 12 per cent, and construction works 3 per cent to total PM emissions (Radian International LLC, 1998).

Monitoring

The PCD currently operates a comprehensive air quality monitoring network consisting of 51 automatic ambient air monitoring stations, of which 17 stations are located in Bangkok, which continuously measure CO, SO_2, Pb, TSP, PM_{10}, NO_x and O_3. In addition, Bangkok also has 21 temporary roadside monitoring stations measuring CO, TSP and PM_{10}. The Bangkok Metropolitan Administration has a mobile unit and ambient air quality and noise monitoring station at Rajathevi district office. Meteorological parameters are also measured at many of the monitoring sites, which increases the capability to analyse and interpret the pollution situation.

Ambient TSP concentrations have fluctuated around the standard of 100 µg/m³ since 1994. Roadside TSP concentrations decreased dramatically from values around 500 µg/m³ in the mid-1990s to 180 µg/m³ in 2004. Since 1997 ambient PM_{10} concentrations in Bangkok have been decreasing from 80 µg/m³ to 40–60 µg/m³ (see Figure 3.1). Roadside PM_{10} values are somewhat higher (approximately

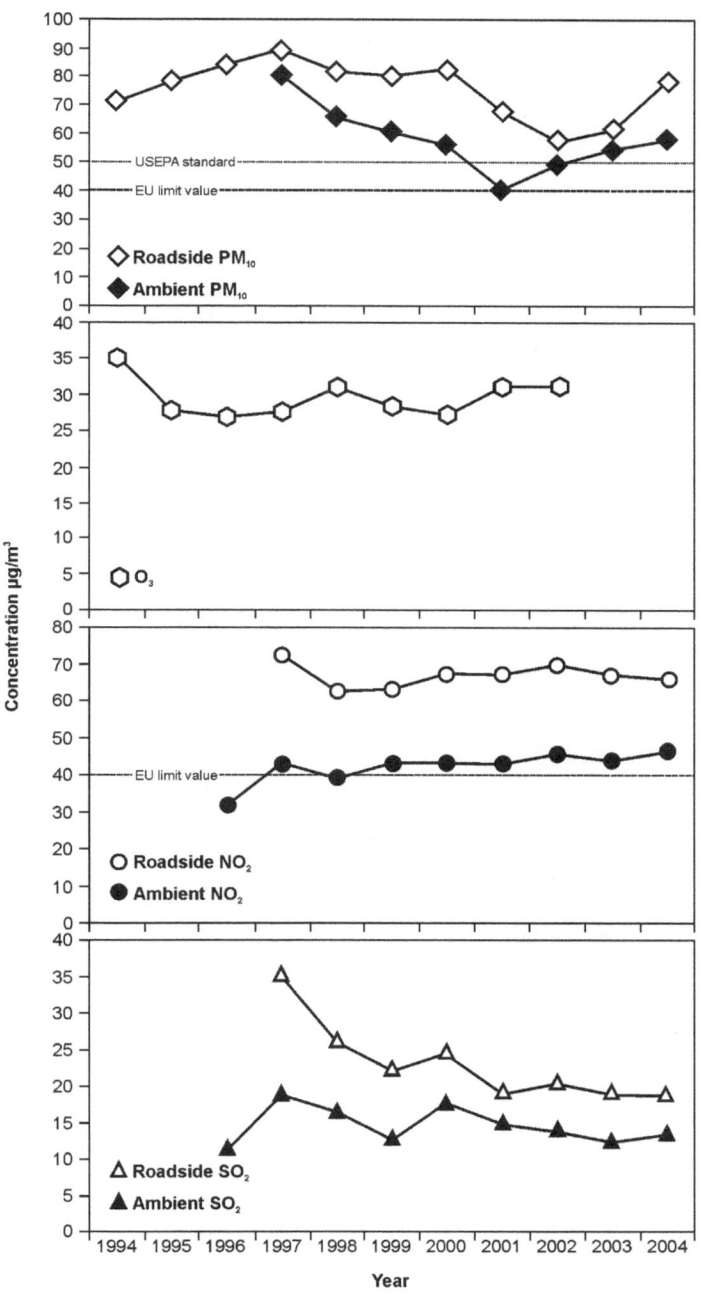

Figure 3.1 *Average annual concentrations of PM$_{10}$, O$_3$, NO$_2$, SO$_2$, Bangkok 1994–2004*

Source: PCD (2005)

10–30 µg/m^3) and follow the general tendency of measurements in non-roadside areas. Annual ambient NO$_2$ concentrations vary between 50 and 60 µg/m^3 since 1994. Roadside NO$_2$ concentrations are higher by 20–25 µg/m^3. Both non-roadside and roadside SO$_2$ concentrations are relatively low, comply with the Thai annual standard and show a decreasing tendency since 1998. Annual CO concentrations show a decreasing tendency, both for roadside and non-roadside sampling stations. Annual O$_3$ concentrations, measured since 1999 fluctuate around 30 µg/m^3 with a stagnant tendency.

Bangkok occasionally experiences photochemical smog due to high levels of O$_3$. In 2003 0.25 per cent of hourly average ambient O$_3$ concentrations exceeded the ambient air quality standard of 100 ppb (196 µg/m^3). The exceedance of the standard mostly occurs in the areas downwind from the centre of Bangkok (Oanh and Zhang, 2004).

Health

Health studies conducted in Bangkok and surrounding areas focus on PM$_{10}$. A 1998 World Bank study on PM showed an association between respiratory diseases and PM$_{10}$ concentrations. The effect of PM$_{10}$ on daily mortality in Bangkok was statistically significant. For example, a 30 µg/m^3 increase of PM$_{10}$ is associated with a 3–5 per cent increase in daily mortality, 7–20 per cent increase in respiratory mortality, and 2–5 per cent increase in cardiovascular mortality. In addition, approximately 4000–5500 premature deaths each year in the Bangkok Metropolitan Region are attributable to short-term exposures to outdoor airborne PM (Radian International LLC, 1998). Studies investigating the impacts of roadside PM (Tamura et al, 2003; Jinsart et al, 2002; Wongsurakiat et al, 1999) showed a significant direct relationship between levels of PM and exposure and the prevalence of respiratory symptoms.

The estimated health impacts and costs of PM$_{10}$ in Bangkok for 2000 amount to US$424 million. This takes into consideration an annual ambient concentration of 64 µg/m^3 affecting 5.7 million people resulting in 1092 excess deaths and 4550 cases of chronic bronchitis (World Bank, 2002).

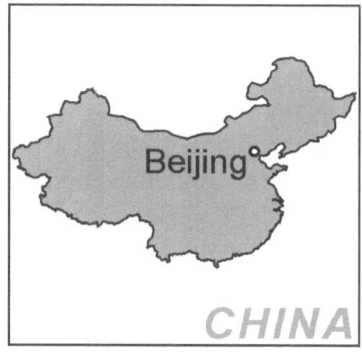

BEIJING

UN estimated population 2003: 10.85 million
UN projected population 2015: 11.06 million
Area: 16,800 km²
Climate: Continental warm summer
Map reference: 39° 52'N, 116° 27'E
Annual mean precipitation: 700 mm
Altitude: 44 m
Annual mean temperature range: –1–31° C

SITUATIONAL ANALYSIS

General

Beijing is the capital of the People's Republic of China. It covers an area of approximately 16,800 km², 38 per cent of which is on flat land. Beijing is approximately 44 metres above sea level and 183 kilometres from the sea. It has a continental warm summer climate. Sandstorms are also a common occurrence especially during spring and winter.

Legislation

The 1989 national Environmental Protection Law is the framework law for the prevention and control of pollution. However, the 1987 Law on Prevention and Control of Atmospheric Pollution sets out regulations for air pollution including environmental impact assessment (Article 11); total emission control and emission licensing (Article 15); prohibition of industrial sources near protected areas (Article 16); banning of equipment causing serious air pollution (Article 19); prohibition of coal washing (Article 24); application of emission standards and controls for fixed and mobile sources (Articles 27, 30, 32).

National ambient air quality standards are categorized into three classes. Designated industrial areas are expected to comply with Class III standards, residential areas with Class II standards, and parks and specially protected areas with Class I standards. Regulation GB3095-1996 requires Beijing to comply with Class II standards. The annual Class II standards for SO_2 (60 µg/m³) and NO_2 (40 µg/m³) are more stringent than those of the USEPA (80 µg/m³ for SO_2; 100 µg/m³ for NO_2), but are lenient or equal to the EU limit values (20 µg/m³ for SO_2; 40 µg/m³ for NO_2). The EU limit for SO_2 protects ecosystems and corresponds to the Class I standard. The Class II standards for 24-hour exposure to SO_2 (150 μg/m^3) are lenient compared with the WHO guideline (20 μg/m^3) and EU limit values (125 μg/m^3) but more stringent than the USEPA standard of 365 μg/m^3. The Class II 1-hour standard for SO_2 (500 μg/m^3) is lenient compared with the EU limit value of 350 μg/m^3 as is the 1-hour standard for NO_2 (240 μg/m^3) compared with the WHO guideline and EU limit values (200 μg/m^3). For O_3 the Chinese standard (160 μg/m^3) is more stringent than the USEPA standard (235 μg/m^3) but more lenient than the EU limit (120 μg/m^3) and the WHO guideline value (100 μg/m^3). The PM_{10} standards for one year and 24 hours (100 μg/m^3, 150 μg/m^3) are more lenient than the EU limit values (40 μg/m^3, 50 μg/m^3), the WHO guideline values (20 μg/m^3, 50 μg/m^3) and the USEPA annual standard (50 μg/m^3) but the Chinese PM_{10} standard for 24 hours equals the USEPA 24 hour standard (150 μg/m^3).

China has implemented a number of regulations to help prevent and control air pollution from mobile sources including standards to regulate exhaust emissions from gasoline and diesel vehicles with normal and full loads. Beijing implements all these standards, with the exception of emission standards for motorcycles as more stringent motorcycle emissions standards were introduced in Beijing before the rest of the country (ADB, 2003).

Both gasoline and diesel fuels distributed in the city need to comply with legal specifications for each fuel type. The sulphur contents of gasoline and diesel, however, are still relatively high at 1000 and 2000 ppm, respectively. These fuel specifications are only Euro I compliant. Gasoline is also monitored for benzene, aromatics and olefins content (ADB, 2003).

Stationary sources of air pollution are also regulated in Beijing. Emissions from thermal power plants; cement plants; coal-burning, oil-burning and gas-fired boilers; coke ovens; industrial kilns and furnaces and even cooking fumes are all subject to emission standards. Specific regulations and standards are also being implemented for solid waste – specifically on incineration of municipal solid wastes and hazardous wastes; storage and disposal sites of solid wastes from general industries; and on pollutants from fly ash use in agriculture (SEPA, 2006).

Emissions

Power plants, boilers and industrial sources are the main source of SO_2 (100 per cent) in Beijing while mobile sources are a main source of CO (74 per cent) (Yu, 2002a). The main source of PM_{10} and $PM_{2.5}$ is industry and power plants (41 and 44 per cent, respectively), mobile-related sources (31 and 33 per cent, respectively) and domestic, construction sources and soil (28 and 23 per cent, respectively) (Zhang et al, 2004).

Monitoring

Beijing currently has 24 stations capable of monitoring Pb, NO_x, O_3, TSP, benzo(a)pyrene (BaP), PM_{10} and $PM_{2.5}$, SO_2 and CO (Yu, 2002b). There are approximately 150 diffusive sampling sites monitoring NO_2 in summer and winter, 100 diffusive sampling sites monitoring O_3 in summer and approximately 100 diffusive sampling sites monitoring SO_2 during winter (Yu, 2002b). In addition, satellite remote sensing and lidar are used for aerosol analysis and modelling. Meteorological parameters such as wind speed, wind direction, temperature and relative humidity are also measured. Figure 3.2 presents the annual means for PM_{10}, NO_x, CO and SO_2.

Although TSP and PM_{10} concentrations have been decreasing in Beijing in the period 1996–2004, they still exceed their respective Class II standards (200 µg/m^3 and 100 µg/m^3, respectively) (BJEPB, 2005). Annual concentrations of NO_x (sum of NO_2 and NO) were approximately 130 µg/m^3 between 1999 and 2003 and decreased in 2004 to a value below the Class II standard (80 µgm^3) of NO_2. This means that NO_2 concentrations in Beijing also complied with the standard in 2004. Since 1998 annual means of CO have shown a decreasing tendency. The annual SO_2 concentrations decreased substantially by more than 50 per cent in the period 1997–2004 and now comply with the Class II standard (60 µg/m^3).

Sun et al (2004) collected aerosol samples of $PM_{2.5}$ and PM_{10} simultaneously at road, industrial and residential sites in Beijing to provide a representative picture of air pollution in the city. The samples were collected in the summer and winter seasons during the period 2002–2003. Twenty-three elements and 15 ions together with organic carbon (OC) and elemental carbon (EC) were analysed systematically in order to characterize the aerosol. $PM_{2.5}$ was the major part of the inhalable particles (PM_{10}), as the ratios of $PM_{2.5}/PM_{10}$ were 0.45–0.48 in summer and 0.52–0.73 in winter. Secondary aerosols (SO_4^{2-}, NO^{3-} and NH_4^+), road dust or/and long-range transported dust, industry and motor vehicles emissions, and coal burning were the major contributors to the airborne particle pollution in Beijing (Sun et al, 2004). Dan et al (2004) demonstrated that coal combustion was a dominant source of carbonaceous species of $PM_{2.5}$ in the urban area of Beijing during winter through comparisons of OC and EC with trace elements (As, Zn, K and Pb) at various sites. Biomass burning, traffic and/or industry emission were the major sources of OC and EC in summer.

Health

In terms of health impacts due to air pollution, several time-series studies have assessed the acute effects of air pollution on mortality, hospital outpatient visits and birth weight in Beijing. The strongest effects of SO_2 and TSP on mortality were consistently seen for respiratory diseases. Studies also revealed increased mortality

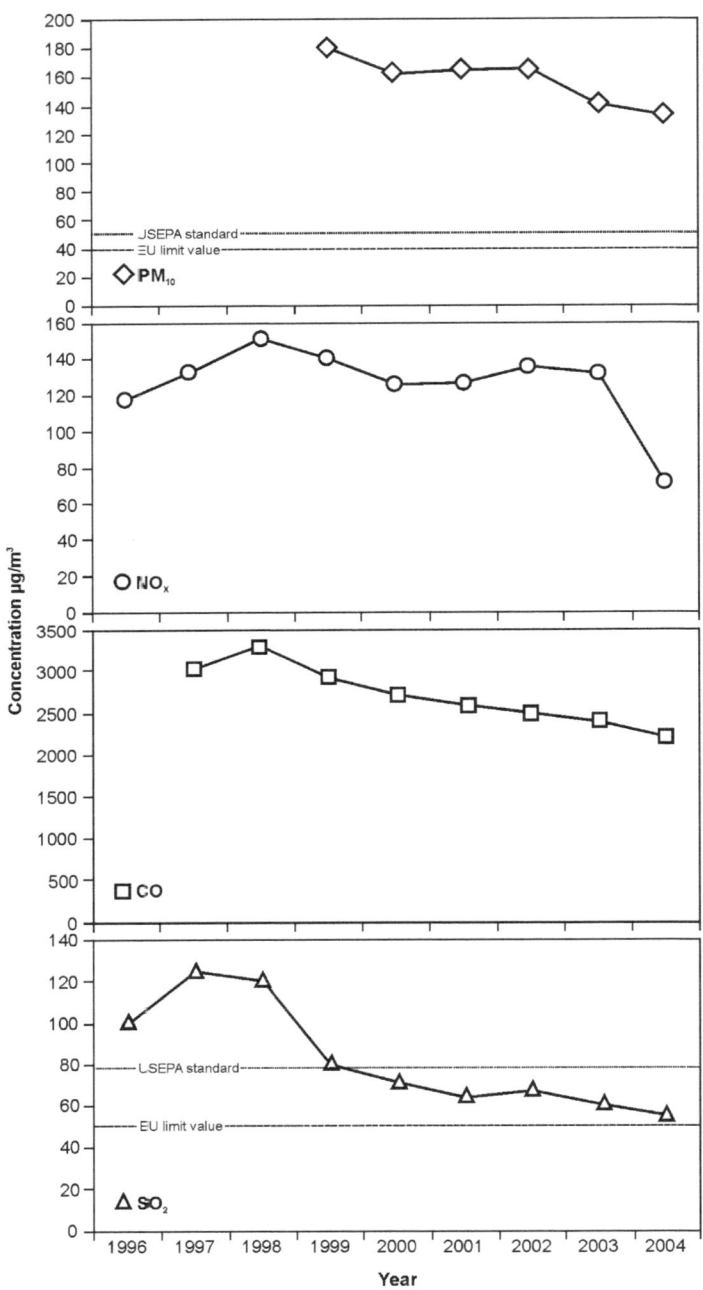

Figure 3.2 *Average annual concentrations of PM_{10}, NO_x, CO, SO_2, Beijing 1996–2004*

Source: BJEPB (2005)

associated with SO_2 pollution levels below the WHO 2000 air quality guideline value of 50 µg/m³ (Xu et al, 1994, 1995a, 1995b; Wang et al 1997; WHO, 2000; Chang et al, 2003). This result is consistent with the recent setting of the 24-hour guideline value to 20 µg/m³ on the basis of European data implying an even lower annual guideline value if any (WHO, 2005).

Kan et al (2005) studied the relationship between daily severe acute respiratory syndrome (SARS) mortality, ambient air pollution and other factors from 25 April to 31 May, 2003 in Beijing. An increase of 10 µg/m³ over a five-day moving average of PM_{10}, SO_2 and NO_2 corresponded to 1.06 (95 per cent confidence interval: 1.00–1.12), 0.74 (0.48–1.13) and 1.22 (1.01–1.48) relative risks of daily SARS mortality, respectively. The authors concluded that daily mortality of SARS might be associated with certain air pollutants in Beijing.

BUSAN

UN estimated population 2003: 3.58 million
UN projected population 2015: 3.40 million
Area: 763.30 km²
Climate: Humid sub-tropical
Map reference: 35° 09'N, 129° 03'E
Annual mean precipitation: 1483.5 mm
Altitude: 69 m
Annual mean temperature range: 2.4–23.9° C

SITUATIONAL ANALYSIS

General

Busan is located on the southeastern tip of the Korean Peninsula and lies approximately 450 km southeast of Seoul. It occupies the basin on the Nakdong River estuary and borders the Blue Sea to the south. It has a total area of approximately 763.30 km² and is 69 metres above sea level. Busan has a humid sub-tropical climate.

Legislation

The 1977 Environment Conservation Law provides the basis for the 1990 Clean Air Conservation Law (CACL) which requires mandatory monitoring of air pollutants, stipulating emission limits and the procedures for the authorization of polluting air emissions from different sources. Under this law, permission must be obtained from the regional environment office for major private and public developments or smaller scale projects that are to be located in areas which do not meet national air quality standards. An environmental impact assessment must be conducted to ensure environmental considerations are included in the development plans to minimize potential impacts.

National air quality standards exist for SO_2, CO, NO_x, O_3, Pb and PM_{10}. However, due to differences in regional characteristics, local governments of each province have the authority to enforce their own municipal ordinances for environmental quality standards. Busan does not have its own air quality standards and follows the national air quality standards. The Korean annual air quality standards for SO_2 (57 µg/m³), NO_2 (94 µg/m³) and lead (0.5 µg/m³) are more stringent than the USEPA standards (78, 100, 1.5 µg/m³, respectively) and, with the exception of lead, less stringent than the EU standards (20, 40, 0.5 µg/m³, respectively). The Korean 8-hour standards for

CO (10,305 µg/m^3) and O$_3$ (120 µg/m^3) are equal or approximately equal to those of the US and the EU. The Korean annual PM$_{10}$ standard of 70 µg/m^3 is less stringent than the corresponding US standard (50 µm/m^3) and EU limit value (40 µg/m^3). The Korean 24-hour standard for PM$_{10}$ (150 µg/m^3) equals that of the US but is three times higher than that of the EU (50 µg/m^3).

The current emission standards for new vehicles powered by gasoline and diesel fuel meet Euro III and 2002 Transitional Low Emission Vehicle (TLEV) standards and will be strengthened to Euro IV and Ultra Low Emission Vehicle (ULEV) standards by 2006. From July 2002 the PM emission standard was lowered from 0.20 g/kWh to 0.10 g/kWh for heavy-duty diesel vehicles. From 2006 the standards will be lowered further to 0.02 g/kWh. Vehicles undergo inspections at least every two years and are subject to mandatory maintenance orders with the possibility of fines if any of the standards are exceeded. Random roadside checks are also conducted with over 200 inspection teams based around Korea. The sulphur content of diesel fuel was lowered to 0.043 per cent (430 ppm) in 2002 and will be further lowered to 0.03 per cent (30 ppm) by 2006.

For stationary sources, the government has implemented policies to reduce the sulphur content in fuels used by industry in order to meet the objective of lowering the annual mean SO$_2$ concentration below the standard of 55 µg/m^3 (20 ppb). To manage air pollutants emitted from manufacturing facilities, emission standards for 18 types of gaseous substances, 9 types of particulate substances and 8 noxious substance types equivalent to EU and Japanese standards were established out of the total of 52 pollutants specified in the CACL (MOE, 2006).

Emissions

On-road mobile sources and ships are responsible for approximately 46 and 19 per cent of SO$_2$ emissions, respectively. Since the amount of fuel sales was used in calculating the emission rate for on-road mobile sources, the actual contribution rate of road mobile sources may be lower than 46 per cent. Approximately 80 per cent of CO is caused by on-road mobile sources. On-road mobile vehicles are the largest source of NO$_x$ emissions at approximately 38 per cent followed by construction machinery (31 per cent). Together these sources account for almost 80 per cent of PM$_{10}$ emissions (Busan, 2002).

Monitoring

Busan comprises a network of 14 monitoring stations located in different areas to include commercial, residential and industrial sectors. The network is intended to provide early warning of high pollution episodes through measurement dispersion

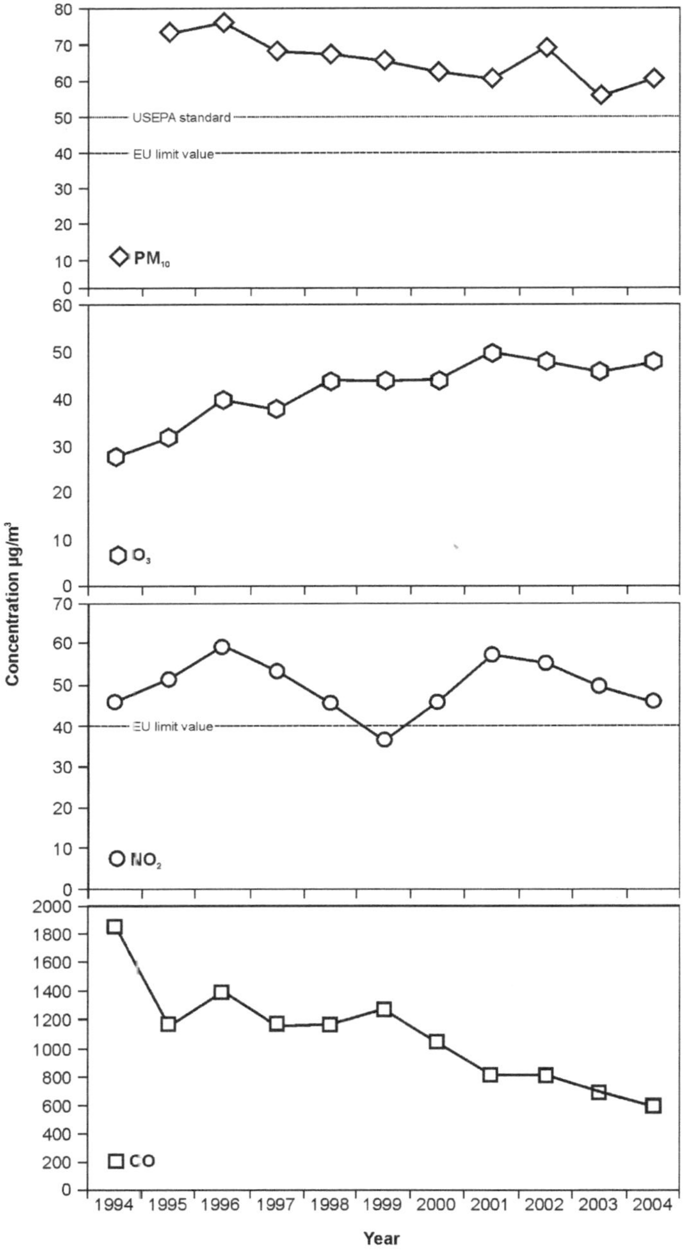

Figure 3.3 *Average annual concentrations of PM$_{10}$, O$_3$, NO$_2$ CO, Busan 1994–2004*

Source: MOE (2006)

modelling, enabling the implementation of emergency air quality control procedures during pollution episodes. The network is fully automated to supply data averaged over a one-hour period. SO_2 levels in Busan have been substantially reduced compared to 1980 levels due to the introduction of cleaner fuels and the regulation of fuel burning systems. Lead values are below the Korean standard and have shown a decreasing tendency since 1998 reflecting the success of lead phase-out in vehicle fuel. TSP has been measured since 1984 while levels of PM_{10} have been monitored since 1995. In January 2000, all instruments were converted to a PM_{10}-based control system. In this process, the measuring sites have been changed and it is difficult to interpret PM_{10} pollution levels from 1995 to 2000 in relation to those after 2000. Figure 3.3 shows the annual means for PM_{10}, O_3, NO_2 and CO.

Since 1995 ambient PM_{10} and CO concentrations have been decreasing, PM_{10} levels now being below the Korean standard of 70 $\mu g/m^3$. NO_2 concentrations increased until 1996 and have now levelled off. The overall NO_x emissions have remained constant over this time period (APMA/KEI, 2002). In contrast, ambient O_3 concentrations have been steadily increasing (Busan, 2001).

Health

In Busan, SO_2 has been identified as a significant predictor for all-cause deaths. An increase of 133 $\mu g/m^3$ (50 ppb) of SO_2 corresponded to a 3.6 per cent increase in excess deaths (95 per cent confidence level: 0.4–6.9 per cent) having controlled for weather conditions and the other pollutants (Lee et al, 2000). Assuming that the mortality in Busan equals that of Korea (520/100,000), approximately 670 excess deaths per year (95 per cent confidence interval: 74–1280) can be associated with an increase of 133 $\mu g/m^3$ in SO_2. This finding is consistent with those of similar studies in Seoul (Lee et al, 1999; Lee and Schwartz, 1999).

The total social costs resulting from air pollution in Busan are approximately 2.5 trillion Won (US$2.5 billion). The costs of PM_{10} account for the largest share at 890 billion Won (US$880 million) (MOE, 2002).

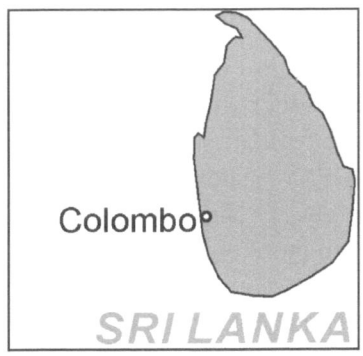

COLOMBO

UN Estimated population 2003: —
UN Projected population 2015: —
Area: 37.29 km²
Climate: Tropical rain forest
Map reference: 6° 56'N, 79 °50'E
Annual mean precipitation: 2400 mm
Altitude: 6 m
Annual mean temperature range: 22–31° C

SITUATIONAL ANALYSIS

General

Colombo is the capital city of Sri Lanka and lies in a coastal area in the lowlands on the southwestern part of the island. The city covers an area of 37.29 km² and is 6 metres above sea level. Colombo has a hot and humid tropical climate.

Legislation

The 1980 National Environmental Act No. 47 was the first comprehensive legislation which covered environmental management and protection in Sri Lanka. In 1981 the Central Environmental Authority (CEA) was established to implement the provisions of this Act. Since then, the CEA has been instrumental in developing the necessary standards relevant for managing air quality in Colombo and for the rest of the country. In 2000 the government approved a national AQM policy to facilitate AQM programmes in Sri Lanka. In July 2001 the Air Resource Management Centre (AirMAC) was formed. Since then AirMAC has been instrumental in improving stakeholder participation in the country. AirMAC is formulating the Clean Air 2009 Action Plan which will need to be approved by the Cabinet of Ministers (Yalegama and Senanayake, 2004). The plan aims to reduce vehicle emissions and improve fuel quality. AirMAC has also focused on accelerating programmes to implement the emissions standards that were scheduled in January 2004 (AirMAC, 2004). The 1994 National Environmental Regulations set out standards for CO, SO_2, NO_2, O_3, TSP and lead.

The Sri Lanka 8-hour CO standard (10,000 µg/m³) is identical to the USEPA standard and the WHO guideline and EU limit value. The 1-hour CO standard (30,000 µg/m³) equals that of the EU and is more stringent than that of the US (40,000

$\mu g/m^3$). For SO_2 and NO_2, Sri Lanka has introduced short-term standards. The 24-hour standard for SO_2 (80 $\mu g/m^3$) is more stringent than the EU limit value (125 $\mu g/m^3$) and the USEPA standard (370 $\mu g/m^3$) but less stringent than the WHO guideline value (20 $\mu g/m^3$). Also the 1-hour standard for SO_2 (200 $\mu g/m^3$) is much more stringent than the EU limit value (350 $\mu g/m^3$). The 1-hour standard for NO_2 (250 $\mu g/m^3$) is lenient compared to WHO guideline and EU limit values (200 $\mu g/m^3$). The Sri Lanka O_3 standard of 200 $\mu g/m^3$ for 1-hour exposure is more stringent than the corresponding USEPA standard (240 $\mu g/m^3$) but lenient compared to the EU limit value (120 $\mu g/m^3$) and the WHO guideline value (100 $\mu g/m^3$). Since Sri Lanka does not have air quality standards for PM_{10} and the EU and the US do not have standards for TSP, a comparison for PM is not possible. However, Sri Lanka uses the US PM_{10} standards to estimate air quality as monitored by PM_{10}. For lead the Sri Lanka standard (0.5 $\mu g/m^3$) equals the WHO guideline and EU limit value and is more stringent than the USEPA standard (1.5 $\mu g/m^3$).

On 1 January 2003 standards for mobile emissions, fuel quality and vehicle importation became effective. Since July 2003, gasoline for high and low octane is unleaded. For sulphur in diesel, the standards require that the maximum allowable sulphur level, now at 0.3 per cent (3000 ppm), be reduced to 0.05 per cent (500 ppm) by 2007 (AirMAC, 2004). Emission standards for stationary sources have not yet been introduced in Sri Lanka.

Emissions

In 1992 the transport sector was the largest contributor to the overall total emissions for all pollutants for TSP (approximately 88 per cent), NO_x (approximately 82 per cent), HC (approximately 100 per cent) and CO (100 per cent) with industry the main contributor to SO_2 emissions (93.5 per cent) (Chandrasiri, 1999).

Emissions from diesel-powered vehicles in Colombo are estimated to account for an annual average exposure to PM_{10} of 25–30 $\mu g/m^3$. Approximately 70 per cent of this is due to primary emissions of PM_{10}, and the remaining 30 per cent is due to the secondary formation of sulphates and, to a lesser extent, nitrates as a result of SO_2 and NO_2 emissions.

Monitoring

In 1997 two fixed air quality monitoring stations were established in Colombo. One station is located at the Fort railway station in the central business district of Colombo, which experiences high volumes of traffic. The second station was based at the meteorological station in the city, which is in a residential area with low traffic.

This second station has now been moved to the premises of the CEA, which is also a low traffic area. These stations were purchased under the World Bank-funded Colombo Urban Transport Project implemented by the Ministry of Transport and handed over to the CEA. The National Building Research Organization (NBRO) operated these stations until 2001; CEA has now taken over their operation. One mobile station is also available and has been used to monitor air quality in areas outside the Colombo Metropolitan Region (CMR), namely Hambantota in the south, Ambewela in the central hill region, Katugastota in the central region, Anuradhapura in the north central region and Puttalam in the northwestern region (Jayaweera, 2001; Malé Declaration, 2000a).

The air quality monitoring stations have been collecting air quality data since 1997. Figure 3.4 presents annual means for PM_{10}, NO_2, CO and SO_2. Ambient PM_{10} levels fluctuate around 80 $\mu g/m^3$ and exceed the USEPA annual standard (50 $\mu g/m^3$) as well as the EU limit value (40 $\mu g/m^3$) and the WHO guideline value (20 $\mu g/m^3$) every year. NO_2 annual concentrations show an overall decreasing tendency since 1997 and are below the USEPA standard (100 $\mu g/m^3$) and in the last three years also below the EU limit and WHO guideline values (40 $\mu g/m^3$). Annual CO concentrations have started to increase in 2003 after a downward tendency over five years. Despite high SO_2 emissions from industrial activities, especially power plants close to Colombo city, the ambient SO_2 levels have been below the adopted limit of 80 $\mu g/m^3$. Since 2000 these levels have decreased to very low values. The low concentrations of most pollutants could be attributed to the dispersion of pollutants due to the relatively unstable atmosphere created by the sea–land breeze (AirMAC, 2004).

Health

Estimates from a World Bank report show that exposure to PM_{10} from diesel vehicles in Colombo can add up to as much as US$ 30 million per year in terms of health damages. The report stated that sulphur level reductions in diesel combined with improved inspection and maintenance of vehicles can result in as much as 70 per cent reduction in health costs (World Bank, 2003).

A study by the National Building Research Organization and the University of Colombo, Faculty of Medicine, found a significant association between ambient air pollution (with respect to SO_2 and NO_x) and acute childhood wheezing episodes in Colombo (Senanayake et al, 1999). According to the WHO, the current level of PM_{10} in Colombo, which is approximately 80 $\mu g/m^3$, is sufficient to cause a 7 per cent increase in daily mortality, 30–35 per cent in bronchitis and other respiratory diseases (WHO, 2000; *The Island*, 2003).

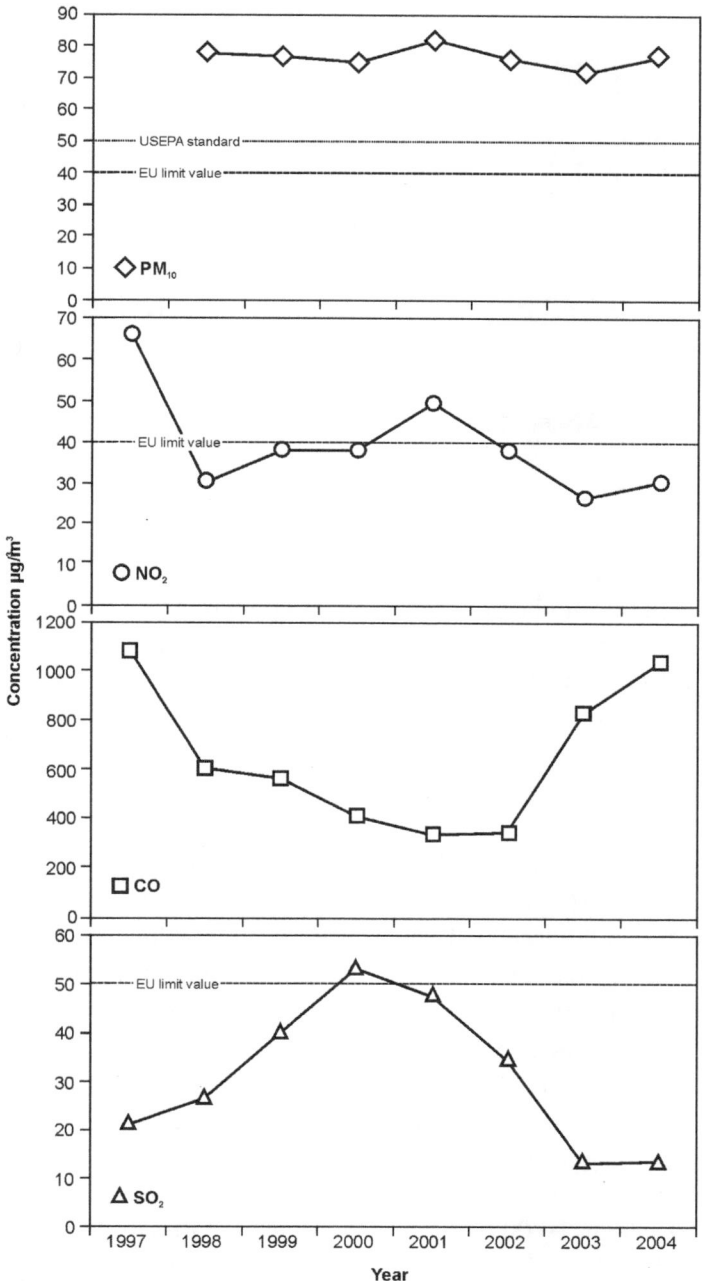

Figure 3.4 *Average annual concentrations of PM_{10}, NO_2, CO, SO_2, Colombo 1998–2004*

Source: Clean Air Sri Lanka Net (2006)

DHAKA

UN estimated population 2003: 11.56 million
UN projected population 2015: 17.91million
Area: 1530 km²
Climate: Tropical rain forest
Map reference: 23° 43'N, 90° 24'E
Annual mean precipitation: 1854 mm
Altitude: 220 m
Annual mean temperature range: 12–35° C

SITUATIONAL ANALYSIS

General

Dhaka is the capital city of Bangladesh. It covers an area of 360 km² and is one of six municipalities in the middle of the Dhaka Metropolitan Area with a total area of 1530 km² (Dhaka City Website, 2006). The city is situated in the populous and flood-prone Ganges-Brahmaputra delta and is 220 metres above sea level. It has a tropical rain forest climate.

Legislation

The 1995 Bangladesh Environmental Conservation Act and the 1997 Environmental Conservation Rules (ECR) are the primary legislation introduced to mitigate air pollution (DOE, 1997; 2002). The ECR outlines the national air quality standards. However, the legislation failed to include appropriate monitoring methods, frequency, averaging times, compliance requirements and other necessary factors, which are important for the enforcement of the air quality standards. A World Bank project proposed a new set of standards equivalent to USEPA standards for PM_{10}, $PM_{2.5}$, NO_2, SO_2, CO, O_3 and to the WHO guideline value for lead (AQMP, 2003a). Standards for CO (1-hour: 40,000 µg/m³), SO_2 (annual: 80 µg/m³; 24-hours: 370 µg/m³) and O_3 (8-hours: 157 µg/m³) tend to be more lenient than the guideline and limit values set by the WHO (CO 1-hour: 30,000 µg/m³; SO_2 annual: –, 24-hours: 20 µg/m³; O_3 8-hours: 100 µg/m³) and the EU (CO 1-hour: 30,000 µg/m³; SO_2 annual: 40 µg/m³, 24-hours: 125 µg/m³; O_3 8-hours: 120 µg/m³) respectively. For NO_2, the annual standard of 100 µg/m³ is more lenient than the WHO guideline and EU limit values (40 µg/m³).

The current motor vehicle emissions standards for Bangladesh are outlined in the ECR. The ECR does not specify if the standards are applicable for new or in-use vehicles or both and for which vehicle type. The 1997 standards were reviewed in 2000 under the Dhaka Urban Transport Project (DUTP) and standards for new vehicles were proposed (AQMP, 2003b). The 2002 Government of Bangladesh and the World Bank Air Quality Management Project (AQMP) recommended standards for new vehicles based on European standards. Euro II standards (termed Bangladesh II) were recommended for gasoline and compressed natural gas (CNG) vehicles excluding 2- and 3-wheelers, while Euro I standards (termed Bangladesh I) were recommended for diesel vehicles. These standards are equivalent to those adopted in Malaysia in 1999, Beijing in 2001 and Taipei, China in 1991.

For in-use vehicles, separate standards were recommended for buses and trucks. The recommended standards are generally higher than those standards from other countries within the region. These were found to be necessary to ensure that a large proportion of the diesel vehicle fleet did not fail the proposed standards. The recommendations included a progressive tightening of the standards over a five-year period to enable government and the private sector to gradually prepare the necessary policies, infrastructure, fuel and technology to meet stricter standards. These standards have not yet been adopted and are published for public debate (DOE, 2005a, 2005b).

In 1997 the Ministry of Energy took action to reduce lead in gasoline due to the heightened awareness of the dangers of lead pollution. The lead content was reduced from 0.8 g/L to an average of 0.4 g/L by blending locally refined leaded gasoline with imported unleaded gasoline. In 1998 low-octane gasoline was made lead-free, but high-octane gasoline still contained 0.4 g/L of lead. In July 1999, due to growing public pressure, the Government of Bangladesh banned the sale of leaded gasoline in the country (DOE, 2001).

At present, Bangladesh does not have a clear strategy to further improve diesel fuel quality in the country. Local sulphur levels remain high at 5000 ppm, while imported diesel contains 2500–5000 ppm sulphur (Chowdhury, 2003).

Emissions

Emission inventories in Dhaka are rare and have been limited to mobile sources, with motor vehicles being responsible for approximately 57 per cent of total emissions of CO from mobiles sources, 14 per cent of TSP, 10 per cent of SO_2, 69 per cent of lead, 12 per cent of NO_x and 36 per cent of HC emissions (World Bank, 1998; ICTP, 2001). Approximately 55 per cent of coarse particulate emissions are attributed to re-suspended soil and motor vehicles. The fine particulates are mostly attributed to motor vehicles (29 per cent) and natural gas/diesel burning (46 per cent) (Biswas et al, 2000).

Monitoring

In 2002 a continuous air quality monitoring station, which meets USEPA Federal Reference Method specifications, was established in the centre of the city as part of a World Bank AQM project. The station monitors NO_x, CO, SO_2, O_3 and non-methane hydrocarbons continuously (Core, 2003; Akhter et al, 2004).

The annual averages of ambient SO_2 for 2002 and 2003 are below 9 μg/m³ for each year. The annual mean concentrations of CO are approximately 1000 μg/m³. The annual averages of ambient NO_2 for 2003 and 2004 are approximately 40 and 50 μg/m³ respectively, below the proposed national and USEPA standard and near to the WHO guideline and EU limit values. Annual averages for PM_{10} are between 110 and 140 μg/m³ and do not comply with the proposed national standard (50 μg/m³). Annual averages of $PM_{2.5}$ were 69 μg/m³ in 2002, 75 μg/m³ in 2003 and 96 μg/m³ in 2004 and exhibit an increasing tendency. These annual averages are above the proposed annual standard of 15 μg/m³ for this compound (AQMP, 2005).

Figure 3.5 shows the monthly mean concentrations of $PM_{2.5}$ and PM_{10}. There is a strong monthly variation for $PM_{2.5}$ and PM_{10} with high peaks during the winter months reaching monthly peak PM_{10} levels at approximately 250 μg/m³.

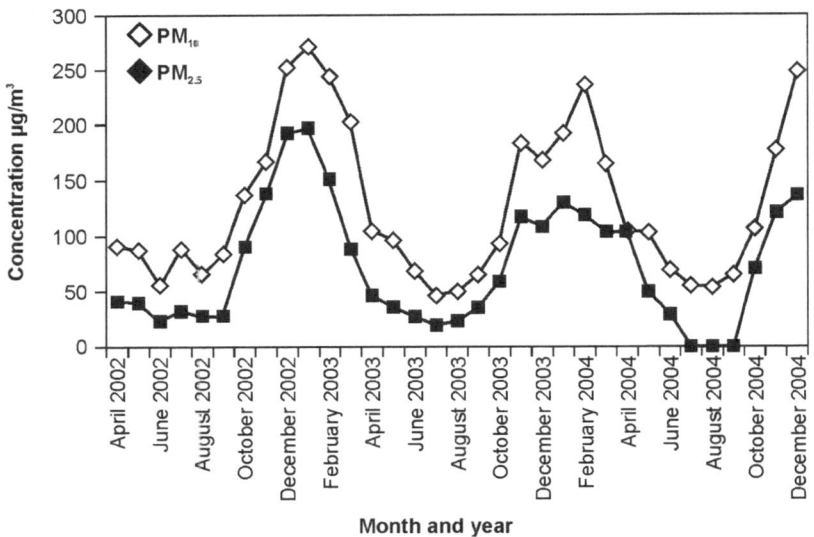

Figure 3.5 *Monthly average of PM concentration at a continuous air monitoring station, Sangsad Bhaban (April 2002 – December 2004)*

Source: AQMP (2005)

Health

Health and environmental impact studies are rare in Dhaka. Air pollution is responsible for an estimated 6000 deaths in Dhaka city each year and, on average, each city dweller spends approximately US$12 per year on medical treatment for pollution-related illnesses (Islam, 2003, corresponding to social costs of US$139 million per year). A few studies have addressed the issue of lead content in blood (Kaiser et al, 2001). VOCs, which are known carcinogens, are a particular problem as emissions of two-stroke auto-rickshaws exceed four to seven times the maximum permissible levels (DOE, 2001). An estimated US$200–800 million per year is the economic cost of sickness and deaths associated with air pollution in Dhaka (World Bank, 1998).

HANOI

UN estimated population 2003: 3.98 million
UN projected population 2015: 5.28 million
Area: 918 km²
Climate: Humid tropical
Map reference: 21° 01′N, 105° 51′E
Annual mean precipitation: 1800 mm
Altitude: 16 m
Annual mean temperature range: 13–33°C

SITUATIONAL ANALYSIS

General

Hanoi is the capital city of the Socialist Republic of Vietnam and is located in a delta along the banks of the Red river. Hanoi has a land area of 918 km² and is 16 metres above sea level. It has a humid tropical climate. The climate of Hanoi is highly influenced by tropical monsoons and is characterized by four seasons.

Legislation

The 1993 Law on Environmental Protection (LEP) provides the basic framework for Vietnam's environmental policy. The 1993 LEP government decree provided guidance on the implementation of the LEP and outlined the responsibilities of the National Environmental Protection Agency. There is still no specific act addressing air quality. However, the Department of Environment of the Ministry of Natural Resources and Environment (MONRE) aims to develop a decree on air pollution control in Vietnam and to adopt a law on clean air (Nguyen, 2004).

The 1995 Vietnamese standards TCVN 5937-1995 outlined national ambient air quality standards for six key pollutants. The standards were updated in 2005 and new standards have been proposed but are not yet enforced (TCVN 5937-2005). Additional standards to control other hazardous air pollutants were also established in 1995 by TCVN 5938-1995. In general, Vietnam standards for SO_2 (24-hours: 125 $\mu g/m^3$), NO_2 (1-hour: 200 $\mu g/m^3$), CO (1-hour: 30,000 $\mu g/m^3$; 8-hours: 10,000 $\mu g/m^3$) and O_3 (1-hour: 120 $\mu g/m^3$) equal the limit values of the EU and are generally more stringent than USEPA standards for SO_2 (24-hours: 370 $\mu g/m^3$), NO_2 (1-hour: 280 $\mu g/ m^3$), CO (1-hour: 40,000 $\mu g/m^3$; 8-hours: 10,170 $\mu g/m^3$) and O_3 (1-hour: 240 $\mu g/m^3$). Only the Vietnam SO_2 24-hour standard is lenient compared with the WHO guideline value of 20 μg/m^3. The 24-hour and annual means for PM$_{10}$, 150 μg/m^3 and 50 μg/m^3 respectively, correspond to the USEPA standards and are less stringent than the EU limit values (50 μg/m^3 and 40 μg/m^3, respectively) and the WHO guideline values (50 μg/m^3 and 20 μg/m^3, respectively). The annual standard for NO$_2$ (40 μg/m^3) equals the EU limit and WHO guideline values and is more stringent than the USEPA standard (100 μg/m^3).

TCVN 6438-2001 provides emission standards for both in-use and newly registered gasoline or diesel vehicles in Vietnam. As with the other major cities in the country, Hanoi is implementing more stringent standards than the rest of the country. Future vehicle emissions standards are also being proposed (ADB, 2003).

Unleaded petrol in Vietnam has been available since July 2001. Benzene, as well as sulphur content, is still higher than that already being implemented in other countries within the region. The sulphur content of unleaded gasoline in the city is 5 per cent compared to 1 per cent in other Asian countries (TCVN 6776-2000). The city offers two types of diesel fuel, one with sulphur content less than 0.5 per cent and another with up to 1 per cent sulphur content (TCVN 5689-1997) (APCEL, 2006).

Vietnam has established a large number of national standards to limit industrial emissions (TCVN 5939-1995; TCVN 5940-1995; TCVN 6991-2001 to TCVN 6996-2001; Le, Van Khoa, 2003). In addition to standards imposed on industries and commercial establishments, a separate regulation is imposed on emissions from medical solid waste (TCVN 6560-1999) (APCEL, 2006).

Emissions

There is no comprehensive emissions inventory for Hanoi. Although a number of studies have been undertaken on the contribution of mobile sources to pollution, data is insufficient to realistically identify the pollution levels caused by each type of fuel and vehicle (ADB, 2002).

A city government-funded survey of 300 factories indicated that only 120 factories out of the 300 are significant air polluters. The data were calculated based on surveyed fuel consumption of every factory in the area although detailed results have not been published (Swisscontact, 2005).

A source apportionment study on coarse and fine particulates was conducted using samples collected in the period January 2001 to July 2002. This showed that soil dust and coal burning is a large contributor of coarse and fine particulates with soil dust contributing 45 per cent and coal fly ash 15 per cent (Hien et al, 2003).

Monitoring

There are currently seven automatic air quality monitoring stations in Hanoi, five of which are fixed and two are mobile. Six of these units were purchased by MONRE within in the period 1999–2002. The last unit was purchased by the Ministry of Construction (MOC) in 2004. These stations are able to measure concentrations of PM_{10}, CO, SO_2, NO_x, O_3 and TSP as well as meteorological conditions. However, the stations are managed and operated separately by different organizations. There is no link between the different stations to compare air quality data. Data exist in various formats. It is therefore difficult to provide an overall assessment of air quality in the city. One mobile station has not been in operation while the mobile station of MOC is used only on a contract basis to assess large construction projects. Passive sampling is also being conducted at a minimum of at least four times a year in various locations (Swisscontact, 2005).

Pollutant concentrations from residential areas as monitored by the then Hanoi Department of Science, Technology and Environment (DOSTE) show that mean annual concentrations of SO_2, NO_2 and TSP levels measured at Van Phuc village are generally much lower than those from the Ly Quoc Su area (Le, Tran Lam, 2003). Annual TSP levels at both sites ranged between 200 and 350 $\mu g/m^3$ and exceeded the standard of 140 $\mu g/m^3$ throughout all years between 1996 and 2002. Furthermore, for the same time period annual means of SO_2 in the Ly Quoc Su area exceeded the respective standard while annual averages of NO_2 just met the standard (Le, Tran Lam, 2003). Figure 3.6 presents more recent monitoring data provided by MONRE.

Annual PM_{10} levels exhibit a decreasing tendency but values are well above the Vietnam standard for all years. Annual O_3 concentrations are below 25 $\mu g/m^3$ and began to level off in 2002. In contrast NO_2 and SO_2 concentrations show an increasing tendency but still comply with their respective standards.

Health

Although epidemiological studies specific to Hanoi do not exist, the high prevalence of allergy and asthma in Hanoi could be related to the degradation of air quality in the city. There are approximately 13 million motorcycles on Vietnam's roads and an increasing number of motor vehicles. The health effects associated with mobile source pollution include asthma, bronchitis and premature deaths, especially in children and the elderly (Vietnam Panorama News Online, 2005).

It is estimated that the annual average cost for treatments of asthma patients in the country is approximately US$301 per person and that approximately 2 million working days and 3 million school days are lost annually due to asthma (Ha, 2004).

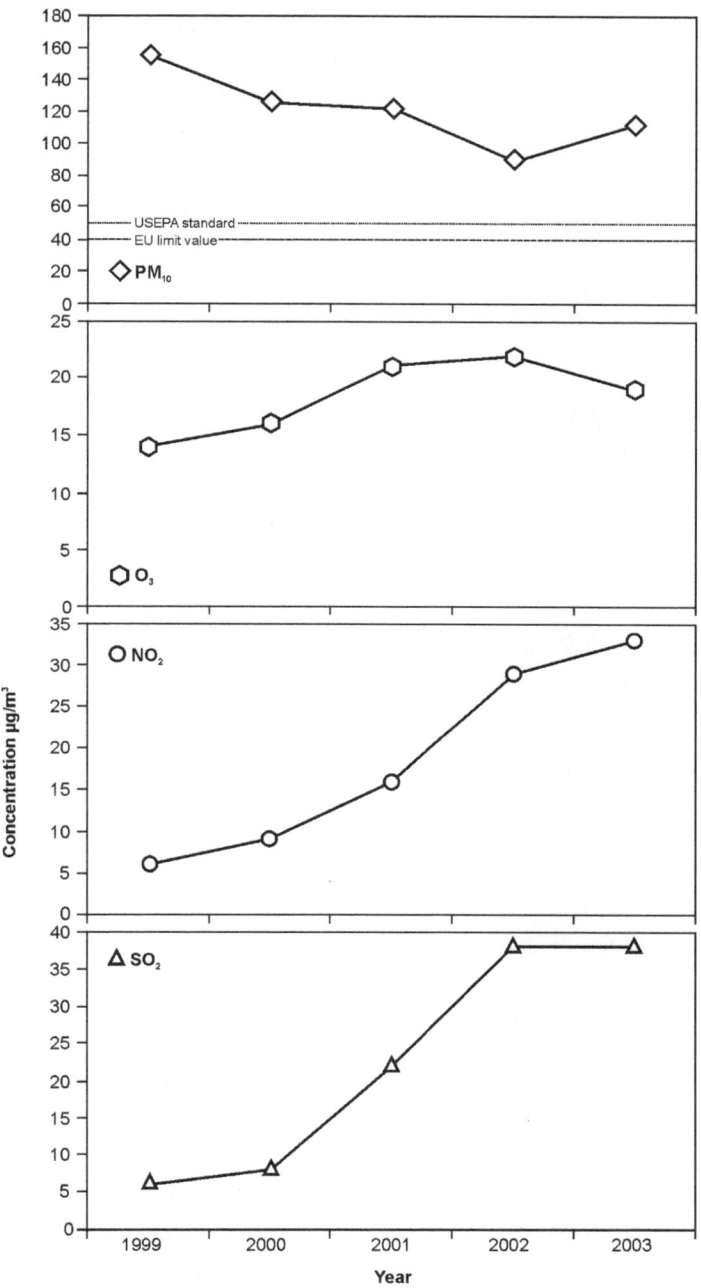

Figure 3.6 *Average annual concentrations of PM$_{10}$, O$_3$, NO$_2$, SO$_2$, Hanoi 1999–2003*

Source: Khaliquzzaman (2005)

HO CHI MINH CITY

UN estimated population 2003: 4.85 million
UN projected population 2015: 6.31 million
Area: 2095 km²
Climate: Tropical savannah
Map reference: 10° 45'N, 106° 42'E
Annual mean precipitation: 1910 mm
Altitude: 2.25 m
Annual mean temperature range: 21–35 °C

SITUATIONAL ANALYSIS

General

Ho Chi Minh City (HCMC) is located in the southeastern part of South Vietnam and is strategically located on the Sai Gon river at the centre of the Cuu Long delta. The city has an area of 2095 km² and is 2.25 metres above mean sea level (Do, 2003). HCMC has a tropical savannah climate.

Legislation

The 1993 Law on Environmental Protection (LEP) provides the basic framework for Vietnam's environmental policy. The 1993 LEP government decree provided guidance on the implementation of the LEP and outlined the responsibilities of the National Environmental Protection Agency. There is still no specific act addressing air quality. However, the Department of Environment of the Ministry of Natural Resources and Environment (MONRE) aims to develop a decree on air pollution control in Vietnam and to adopt a law on clean air (Nguyen, 2004).

The 1995 Vietnamese standards TCVN 5937-1995 outlined national ambient air quality standards for six key pollutants. The standards were updated in 2005 and new standards have been proposed but are not yet enforced (TCVN 5937-2005). Additional standards to control other hazardous air pollutants were also established in 1995 by TCVN 5938-1995. In general, Vietnam standards for SO_2 (24-hours: 125 $\mu g/m^3$), NO_2 (1-hour: 200 $\mu g/m^3$), CO (1-hour: 30,000 $\mu g/m^3$; 8-hours: 10,000 $\mu g/m^3$) and O_3 (1-hour: 120 $\mu g/m^3$) equal the limit values of the EU and are generally more stringent than USEPA standards for SO_2 (24-hours: 370 $\mu g/m^3$), NO_2 (1-hour: 280 $\mu g/ m^3$), CO (1-hour: 40,000 $\mu g/m^3$; 8 hours: 10,170 $\mu g/m^3$) and O_3 (1-hour: 240 $\mu g/m^3$). Only the Vietnam SO_2 24-hour standard is lenient compared with the WHO guideline value of 20 $\mu g/m^3$. The 24-hour and annual means for PM_{10}, 150 $\mu g/m^3$ and 50 $\mu g/m^3$ respectively, correspond to the USEPA standards and are less stringent than the EU limit values (50 $\mu g/m^3$ and 40 $\mu g/m^3$, respectively) and the WHO guideline values (50 $\mu g/m^3$ and 20 $\mu g/m^3$, respectively). The annual standard for NO_2 (40 $\mu g/m^3$) equals the EU limit and WHO guideline values and is more stringent than the USEPA standard (100 $\mu g/m^3$).

Unleaded petrol in Vietnam has been available since July 2001. Benzene, as well as sulphur content, is still higher than that already being implemented in other countries within the region. As with the rest of Vietnam, most of HCMC's transport means rely only on either gasoline or diesel fuels, which are mainly imported from other countries. Standard TCVN 5689-1997, requires Vietnam's diesel fuel to have sulphur content no higher than 1 per cent of its weight. In fulfilment of this requirement, Petrolimex, the country's largest petrol importer, imports diesel with sulphur content no higher than 0.5 per cent of its weight. By virtue of TCVN 6704-2000, the gasoline being imported to and used in the country can only have a maximum of 0.013 g/L lead content. Since July 2001 unleaded gasoline has been introduced with the aim to reduce ambient lead concentrations to below the WHO guideline value (Vietnam Register, 2002). Mandatory emission standards for new automobiles and motorcycles are laid down in TCVN 6438: 2001 – Air quality – Road vehicles emission – maximum permitted limit. Voluntary emission standards for motorcycles and mopeds are promulgated in TCVN 7357-2003 and TCVN 7358-2003 (APCEL, 2006).

Vehicles failing to meet the emission standards set by law can be suspended from operation with their drivers subject to a fine of VND 500,000 (US$31). Inspection of in-use vehicles in the country began in August 1995 and since then all motor vehicles have been required to undergo periodic inspections before they are given appropriate inspection certificates and allowed on the roads. Inspection certificates are given out only after the vehicles pass all 55 inspection test items. Compulsory emissions testing in the country began to be implemented in August 1999 (Vietnam Register, 2002).

Vietnam has established a large number of national standards to limit industrial emissions (TCVN 5939-1995; TCVN 5940-1995; TCVN 6991-2001 to TCVN 6996-2001) (APCEL, 2006; Le, Van Khoa, 2003).

Emissions

As Vietnam's industrial and economic hub, the more dominant emission sources in HCMC are industry and transport. Industrial plants contribute largely to SO_2 (92 per cent) and CO_2 emissions while mobile emissions are the main emitters of CO (84 per cent), NO_x (61 per cent) and HC (94 per cent) (Vietnam Register, 2002). Motorcycles are a main contributor to emissions of CO, hydrocarbons and volatile organic compounds while trucks are a major contributor to vehicle-related emissions of SO_2 and NO_2. Both trucks and motorcycles are the major emitters of vehicle-related CO_2 (Dam, 2004).

Monitoring

The automatic air quality monitoring network in HCMC has been operational since 1992. Until 2002 the network was composed of four residential and three roadside monitoring stations. The residential monitoring stations measured PM, SO_2 and NO_2 while the roadside stations measured PM, NO_2, lead and noise. In November 2002, three automatic residential and two automatic roadside stations were also installed with support from Norwegian International Development Agency (NORAD) increasing the total number of automatic stations to nine.

Generally, air quality has been improving for TSP and SO_2. In 1998 TSP concentration at roadside stations decreased from levels between 500 and 2000 $\mu g/m^3$ to levels of approximately 500 $\mu g/m^3$ in 2004. However, these levels exceed the annual TSP standard of 140 $\mu g/m^3$. The reductions in the measured levels can be attributed to changes in traffic pattern and improvements to the road surface (Sivertsen et al, 2005). Annual SO_2 concentrations have decreased since 2002 for three of the four long-established sites. Figure 3.7 presents the annual levels for PM_{10}, O_3, NO_2 and CO at roadside and residential stations. Roadside and residential PM_{10} concentrations fluctuate between 60 and 120 $\mu g/m^3$ and exceed the 50 $\mu g/m^3$ standard. O_3 shows a slight increasing tendency at roadside stations and is fairly stagnant at residential stations but is not considered to be a pollutant of key concern in the city as hourly ozone values for the year 2004 have not exceeded the 1-hour standard more than 0.20 per cent of the time. Annual concentrations of NO_2 at roadside stations are slightly increasing and exceed the standard (40 $\mu g/m^3$) and are stagnant and below the standard at residential stations. CO levels are fairly constant throughout the years both at roadside and residential stations.

Health

In 1995 the health effects of air pollution on traffic police officers were studied by the HCMC Labour Protection Unit. The average tuberculosis infection rate of HCMC residents is 0.075 per cent. However, police officers exposed to high levels of air pollution have a higher infection rate of 2.9 per cent. Police officers were also observed to have higher incidences of ear, nose and throat infections (Dang, 1995).

Statistics on the most common illnesses in HCMC hospitals also show respiratory illnesses were consistently at the top from 1996 to 2000. The high occurrence of respiratory-related illnesses may suggest that polluted air in HCMC has important implications for public health (Department of Public Health, 2001).

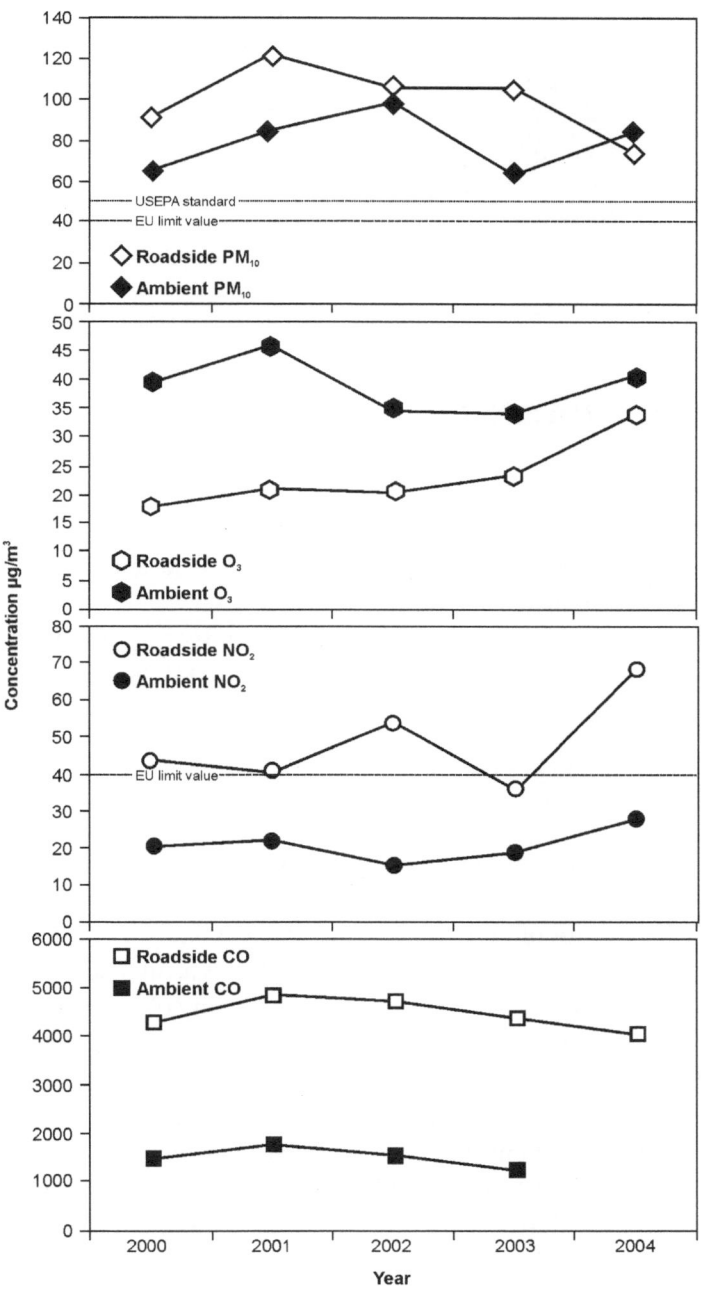

Figure 3.7 *Average annual concentrations of PM_{10}, O_3, NO_2, CO, Ho Chi Minh City 2000–2004*

Source: HEPA (2006)

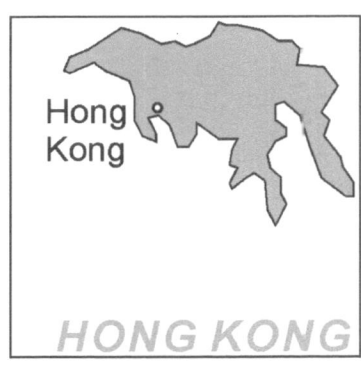

HONG KONG

UN estimated population 2002: 7.05 million
UN projected population 2010: 7.87 million
Area: 1102 km²
Climate: Sub-tropical
Map reference: 22° 16′N, 114° 11′E
Annual mean precipitation: 2200 mm
Altitude: 0 m
Annual mean temperature range: 13–31°C

SITUATIONAL ANALYSIS

General

Hong Kong is located at the southeastern part of China and covers Hong Kong Island, the Kowloon peninsula and the 235 outlying islands called the New Territories totalling an area of 1102 km². Since 1997 it has been a Special Administrative Region of the People's Republic of China (Mottershead, 2002). Its lowest point is at mean sea level and it has a sub-tropical climate.

Legislation

The 1983 Air Pollution Control Ordinance (APCO) provides the statutory framework that enabled the establishment of air quality objectives and subsidiary regulations which address the control of emissions from vehicle exhausts, construction dust and major stationary sources. Under the APCO, Hong Kong is divided into ten air control zones all of which are targeting a uniform set of air quality objectives (AQOs). APCO is made operational by various Technical Memoranda which establish the AQOs for each zone (Mottershead, 2002).

Although mainland China has national ambient air quality standards, Hong Kong enforces its own locally established AQOs. The Hong Kong ambient AQOs are a set of standards for TSP, PM_{10}, SO_2, NO_2, CO, O_3 and lead for varying averaging times. For most compounds and averaging times the AQOs are more lenient than Chinese Class II standards except for TSP, lead and long-term PM_{10}. Hong Kong SO_2 AQOs for 24 hours (350 µg/m³) and one year (80 µg/m³) are more lenient than the WHO guideline (SO_2 24-hour: 20 µg/m³) and EU limit values (SO_2 24-hour: 125 µg/m³ and 1 year: 20 µg/m³) but slightly more stringent than or equal to USEPA standards (SO_2 24-hour: 365 µg/m³ and 1-year 78 µg/m³). For 1-hour exposure Hong Kong's SO_2 AQO (800 µg/m³) is less lenient than the EU limit value of 350 µg/m³. A similar relationship exists between the Hong Kong AQO for NO_2 (1-hour: 300 µg/m³, 1-year:

80 µg/m³) and the WHO guideline and EU limit values (1-hour: 200 µg/m³, 1-year: 40 µg/m³) Hong Kong's CO AQOs are identical to WHO and EU for 1-hour (30,000 µg/m³) and 8-hours (10,000 µg/m³) averaging times, while the 1-hour standard is more stringent than the USEPA standard (40,000 µg/m³). For O_3 the Hong Kong 1-hour AQO (240 µg/m³) corresponds to that of the US. For PM_{10} the AQOs for 24-hours (180 µg/m³) and 1-year (55 µg/m³) are more lenient than those of the EU (24-hours: 50 µg/m³, 1-year: 40 µg/m³) and the guideline values of the WHO (24-hours: 50 µg/m³, 1-year: 20 µg/m³). The AQO for lead (1.5 µg/m³) is equal to the USEPA standard and is more lenient than WHO and EU values (0.5 µg/m³).

In July 2000, Hong Kong became the first city in Asia to introduce ultra-low sulphur diesel, which has a sulphur content of less than 0.005 per cent, for use in motor vehicles. From April 2002, it became the statutory standard for motor diesel (Hong Kong, 2003).

With regard to emission standards, Hong Kong has adopted Euro III emission standards since January 2001 for all newly registered vehicles except newly registered private diesel cars which must meet emission standards which are more stringent than the Euro III standards. Cleaner fuels and tighter emission standards introduced in the past few years have significantly reduced motor vehicle pollution (Hong Kong, 2003).

Hong Kong has promoted cleaner fuels and vehicles by replacing diesel vehicles with liquid petroleum gas (LPG) vehicles. This incentive programme was completed at the end of 2003. A similar scheme was launched in August 2002 to encourage the early replacement of diesel light buses with LPG or electric light buses. A programme to retrofit pre-Euro diesel light vehicles with particulate reduction devices was completed in October 2001. A similar retrofit programme for pre-Euro diesel heavy vehicles began in December 2002. A new regulation to mandate the installation for pre-Euro diesel light vehicles of up to four tonnes was introduced in December 2003 (Hong Kong, 2003).

Another motor vehicle emission control strategy is to tighten the control against smoky vehicles. All vehicles reported under the scheme must be smoke tested by an advanced test method using a chassis dynamometer at designated vehicle emission testing centres to confirm that the vehicle owners have rectified the smoke defects. The government has adopted a policy to promote mass transit systems that are pollution-free at street level.

With regards to industrial facilities, the Environmental Protection Department (EPD) operates a range of controls under the APCO and its subsidiary regulations, including licensing of some large industrial facilities and specific controls on furnace and chimney installations, dark smoke emissions, fuel quality, open burning, dust emissions from construction works, emissions from petrol filling stations and perchloroethylene emissions from dry-cleaning facilities (EPD, 2004a).

Emissions

In Hong Kong, the public electricity generation sector is a major contributor to emissions of SO_2 (90 per cent), NO_x (76 per cent) and PM (37 per cent). The road transport sector is a major source of PM (38 per cent) and CO (85 per cent) while non-combustion is a major emitter of NMVOC emissions (79 per cent). This includes dust from road construction and other industries which can be transported to locations far away from the source. Overall there is a downward trend in the emissions of all these pollutants (EPD, 2004a).

A serious problem exists with respect to transboundary pollution from the Pearl River Delta (PRD) affecting air quality in Hong Kong. The PRD emits approximately 95 per cent of PM_{10}, 88 per cent of VOC, 87 per cent of SO_2 and 80 per cent of NO_2, while Hong Kong is only responsible for 5 per cent, 12 per cent, 13 per cent and 20 per cent of emissions for these compounds, respectively (Civic Exchange, 2004).

Monitoring

Hong Kong's air quality monitoring network consists of 14 fixed monitoring stations, 11 of which are used to monitor ambient air quality while three are used to monitor roadside air pollution. The Air Services Group of the EPD is responsible for the operation, maintenance and management of these 14 fixed stations. The pollutants continuously monitored are SO_2, NO_x, NO_2, O_3, CO and PM_{10} (respirable suspended particulate). Figure 3.8 shows the levels of PM_{10}, O_3, NO_2 and SO_2 for the period 1994–2004 (EPD, 2005).

While TSP in urban and new town areas have been steadily declining since 1995, annual concentrations of PM_{10} have been stagnant and complied with the AQO (55 $\mu g/m^3$) between 1998 and 2002. After 2002 slight increases of annual values have been noted. Roadside concentrations of PM_{10} continue to exceed the AQO despite decreasing by 13 per cent in 2003 compared with 1999 (EPD, 2004b). The annual average of PM_{10} for urban stations decreased from 1992 to 2000 but the trend has levelled off in subsequent years. Annual levels of O_3 show a rising tendency since 2000 while annual average NO_2 concentrations have remained relatively constant for the past decade. Annual average O_3 for urban stations in 2003 (34 $\mu g/m^3$) was 70 per cent higher than the 1993 value (20 $\mu g/m^3$), with rural areas experiencing higher concentrations compared to urban areas. In 2002 roadside concentrations of NO_2 were reduced by 16 per cent compared to 1999, further reflecting the success of additional vehicle emission control measures (EPD, 2004b). NO_2 concentrations have been observed to be lower than the annual air quality objective (80 $\mu g/m^3$).

The introduction of low-sulphur fuel content contributed to the reduction in SO_2 concentrations, which have remained below the annual Hong Kong AQO of 80

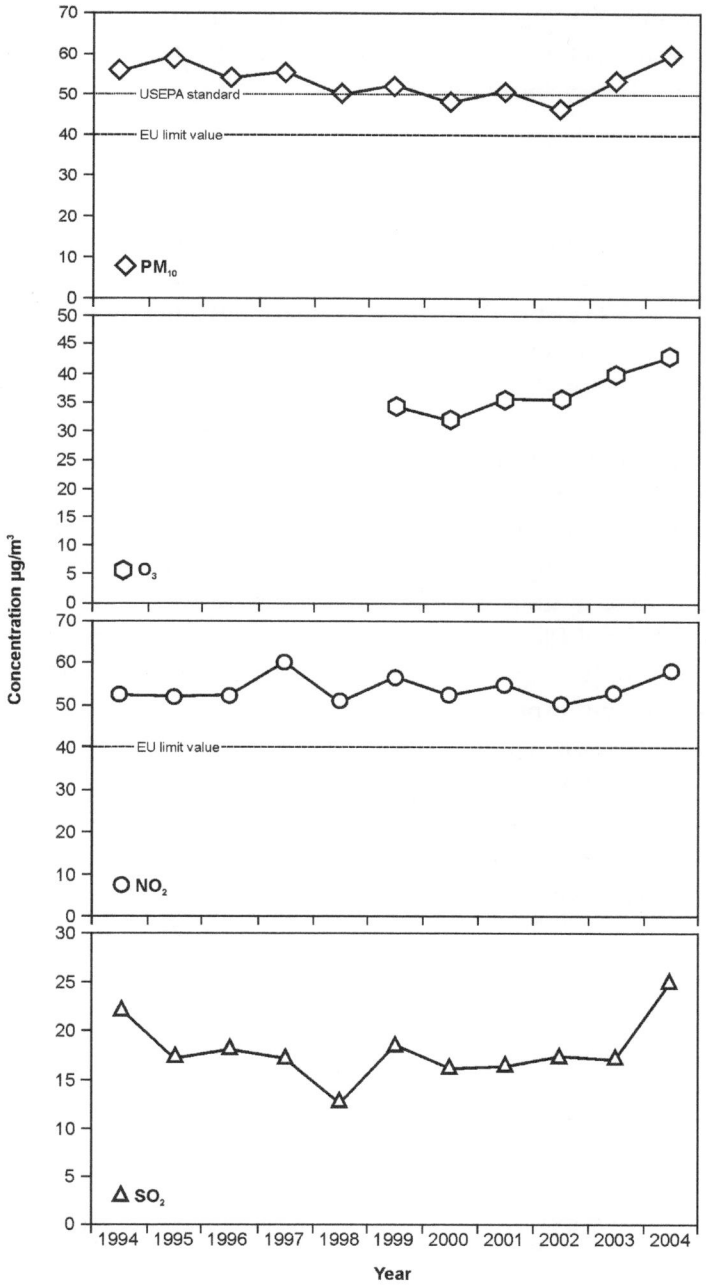

Figure 3.8 *Average annual concentrations of PM$_{10}$, O$_3$, NO$_2$, SO$_2$, Hong Kong 1994–2004*

Source: EPD (2005)

μg/m^3. The average SO$_2$ roadside concentration in 2003 (16 μg/m^3) fell by 43 per cent compared with the 2000 value (28 μg/m^3) (EPD, 2004b). CO concentrations have remained relatively low and below the AQOs.

Health

A number of studies have investigated the effects of air pollution on human health in Hong Kong. Wong et al (1999) investigated short-term effects of ambient air concentrations of NO$_2$, SO$_2$, O$_3$ and PM$_{10}$ on hospital admissions for cardiovascular and respiratory diseases in Hong Kong. Significant associations were found between hospital admissions for all respiratory diseases, all cardiovascular diseases, chronic obstructive pulmonary diseases, heart failure and concentrations of all four pollutants. Admissions for asthma, pneumonia and influenza were significantly associated with NO$_2$, O$_3$ and PM$_{10}$. The relative risk for admissions for respiratory disease for the four pollutants ranged from 1.013 (for SO$_2$) to 1.022 (for O$_3$), and for admissions for cardiovascular disease from 1.006 (for PM$_{10}$) to 1.016 (for SO$_2$). Those aged 65 years old and above were at greater risk. Significant positive interactions were detected between NO$_2$, O$_3$ and PM$_{10}$, and between O$_3$ and winter months. However, Wong et al (2001) found only marginally significant associations between daily mortality due to cardiovascular and respiratory diseases and PM$_{10}$ levels. For hospital admissions due to cardiac diseases, significant associations were observed with PM$_{10}$, SO$_2$ and NO$_2$ (Wong et al, 2002).

In 2000 the direct medical costs, the amount spent on dealing with the excess health problems due to the annual average air pollution level, amounted to approximately US$167 million spending on health care (Wong, 2002).

JAKARTA

UN estimated population 2000: 12.30 million
UN projected population 2010: 17.50 million
Area: 661.52 km²
Climate: Tropical rain forest
Map reference: 6° 07'S, 106° 48'E
Annual mean precipitation: 1760 mm
Altitude: 7 m
Annual mean temperature range: 23–31°C

SITUATIONAL ANALYSIS

General

Jakarta is the capital city of the Republic of Indonesia. It is located on a flat terrain on the northern coast of Java Island near the Ciliwung river, and covers an area of 661.52 km². It is approximately 7 metres above sea level and has a tropical rain forest climate (ADB, 2002; World Bank, 1997).

Legislation

The 1997 Government Act No. 23 on Environmental Management and the 1999 Government Regulation No. 44 on Air Pollution Control are the main acts controlling air pollution in Indonesia. The 1999 Act sets out the mandate for setting up standards and acceptable practices in air pollution control for stationary and mobile sources (ADB, 2002). The 1999 Government Decree No. 41 established national ambient air quality standards.

Under the 2001 DKI Jakarta Governor's Decree No. 551 the city of Jakarta has its own set of standards which are stricter than certain national standards for SO_2, NO_2, O_3 and CO. The standard for SO_2 for 24-hour exposure (260 µg/m³) is more lenient compared to WHO guideline (20 µg/m³) and EU limit values (125 µg/m³) but more stringent than the USEPA standard (370 µg/m³). The annual mean for SO_2 (60 µg/m³) is also more stringent than the USEPA standard (80 µg/m³). The SO_2 1-hour standard of 900 µg/m³ is more than twice the EU limit value of 350 µg/m³. The NO_2 1-hour (400 µg/m³) and annual standards (60 µg/m³) for Jakarta are also more lenient than WHO guideline and EU limit values (200 µg/m³ and 40 µg/m³, respectively). The O_3 standard for 1-hour exposure (200 µg/m³) is more stringent than the USEPA standard (240 µg/m³), but more lenient towards the EU limit (120 µg/m³) and the WHO guideline value (100 µg/m³). The 24-hour PM_{10} standard (150 µg/m³) is lenient compared to the EU limit and the WHO guideline value (50 µg/m³) and equals the USEPA standard. The CO 1-hour exposure standard (26,000 µg/m³) is stricter than WHO, USEPA and EU values.

For mobile emission standards the MOE has issued a notification of the enforcement of Euro II type approval standards taking effect in 2005. The standards were decreed by MOE in October 2003 (Tamin and Rachmatunisa, 2004). In Jakarta the standards for in-use vehicles are stricter than those of the national government as notified under the 2001 Governor of DKI Jakarta Decree No. 1041. This legislation was implemented in order to support the development of the inspection and maintenance system for private passenger vehicles in Jakarta. The local standards, unlike the national, specify the emissions standard based on the production year and vehicle technology (ADB, 2002).

In 2000 the Local Government of Jakarta introduced an Inspection and Maintenance Programme (I&M) which requires all vehicles to comply with emission quality standards and stipulates that the inspection and maintenance system will use a decentralized system, involving the private sector with the local government as the facilitator (Tamin and Rachmatunisa, 2004; World Bank, 2003).

Emissions

In Jakarta mobile sources account for approximately 80 per cent of the total SO_2 emissions and 87 per cent of the total NO_2 emissions, while stationary sources (industry and power generators, utilities and household) account for 91 per cent of PM emissions compared to approximately 8 per cent emitted by mobile sources (Wirahadikusumah, 2002). Passenger cars and motorcycles are each responsible for approximately 40 per cent of HC emissions. More then 50 per cent of NO_x is emitted from private motor vehicles and approximately 30 per cent from buses. Private motor vehicles, buses and trucks are responsible for an equal share of sulphur oxides and PM_{10} emissions (Wirahadikusumah, 2002).

Monitoring

An air quality monitoring system has been operational in Jakarta since 2000. The Indonesian Ministry of Environment has one mobile station and the Jakarta Environmental Management Board (DKI Jakarta) has five fixed continuous monitoring stations linked to the air quality monitoring station network (Tamin and Rachmatunisa, 2004). Pollutants monitored include TSP, PM_{10}, NO_2, SO_2, O_3 and lead.

Data for SO_2 are considered unreliable because of their large variance. Annual values for TSP exceeded the standard (230 µg/m³) in 2001. Data for 2002 to the present are not available. In the 1990s lead concentrations were below the WHO guideline value of 0.5 µg/m³. However, in 2000 this air quality guideline was exceeded (World Bank, 2003). From 2002–2003 there was a substantial decrease in the average lead concentrations at various sampling locations in Jakarta city due mainly to the phasing out of lead in vehicle fuel in 2001 (Nugroho, 2003).

Figure 3.9 presents the air pollutant levels in residential areas for PM_{10}, O_3, NO_2 and CO in 2000–2004. Indonesia has not adopted an annual standard for PM_{10}, therefore compliance of the data with a national standard cannot be undertaken. Annual PM_{10} levels show a rising tendency with values exceeding the USEPA standard (50 µg/m³) and approaching 100 µg/m³ in 2004. Ambient concentrations of O_3 have an increasing tendency to exceed the national standard (30 µg/m³) in all years with concentrations double this limit in 2003. In 2002 annual concentrations of NO_2 exceeded the Indonesian air quality standard (60 µg/m³) while annual CO concentrations have remained relatively low and stagnant.

Health

Health impact studies related to outdoor air pollution are rare in Jakarta. A survey of the respiratory health impacts on mothers and children in Jakarta and Tangerang related prevalence of various respiratory symptoms with NO_2 concentrations. An estimate of the costs of workdays lost for adults amounted to approximately Rupiah 15,639–18,165 (US$6.80–7.90) per episode and adult in Central Jakarta (Duki et al, 2003).

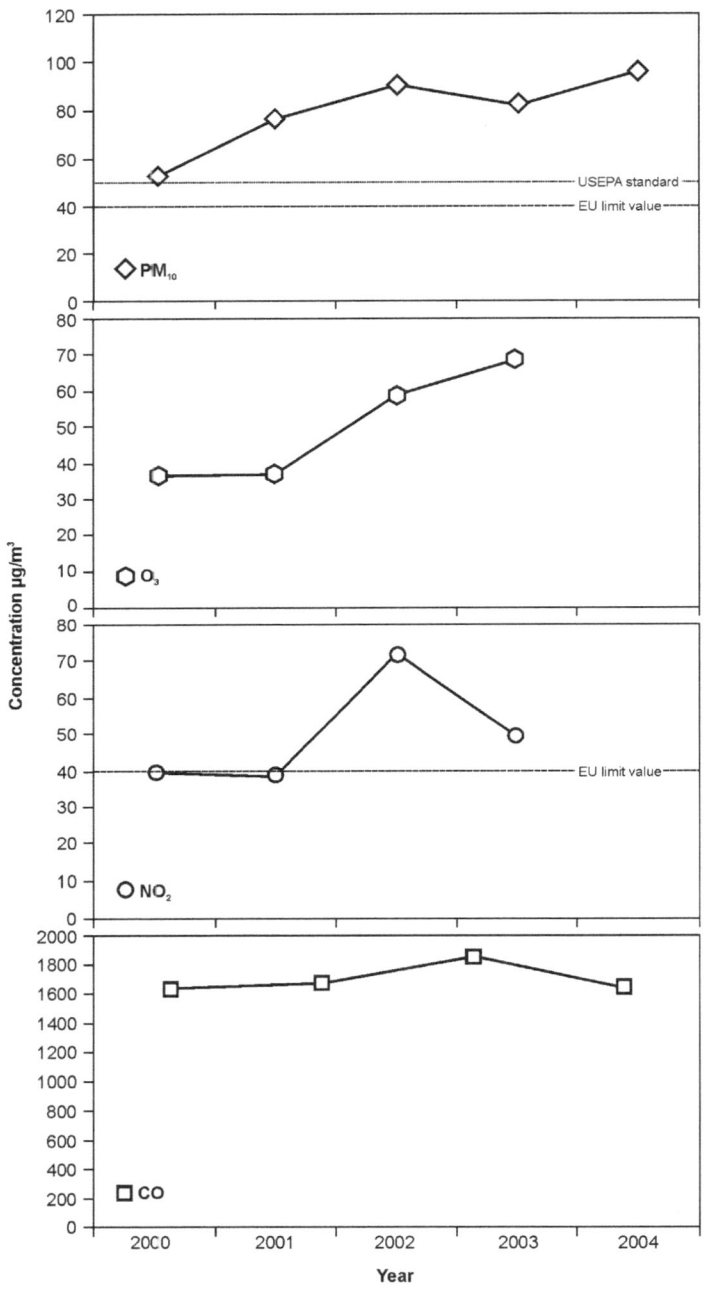

Figure 3.9 *Average annual concentrations of PM$_{10}$, O$_3$, NO$_2$, CO, Jakarta 2000–2004*

Source: MoE (2005)

KATHMANDU

UN estimated population 2003: —
UN projected population 2015: —
Area: 50.67 km²
Climate: Humid sub-tropical
Map reference: 27° 42'N, 85° 18'E
Annual mean precipitation: 1307 mm
Altitude: 1300 m
Annual mean temperature range: 2–30° C

SITUATIONAL ANALYSIS

General

Kathmandu is the capital city of Nepal and is the largest urbanized area in the country covering 50.67 km². It lies in a valley and is 1300 metres above sea level. It is surrounded by hills ranging from 2100 to 3132 metres. Kathmandu has a humid sub-tropical climate. Due to the valley's topography, the occurrence of calm low wind-speed conditions is common in this region, and the formation of temperature inversions, especially in winter, encourages air pollutants to accumulate at high concentrations (Shrestha and Raut, 2002).

Legislation

The legislative framework to control air pollution in Nepal is fragmented over 16 different acts and by-laws. The 1997 Environment Protection Act and Environment Protection Rules act as a framework law on the environment (MOPE, 2005). The Act covers environment conservation; pollution control and prevention; conservation of natural heritage sites; operation of environmental funds; additional incentives to minimize pollution; and compensation for environmental damage.

In June 2003 the Ministry of Population and Environment set national ambient air quality standards for TSP, PM_{10}, SO_2, NO_2, CO, lead and benzene (MOPE, 2005). The annual air quality standards for NO_2 and lead coincide with the WHO guideline and EU limit values (40 µg/m³ and 0.5 µg/m³, respectively) as does the 8-hour standard for CO (10,000 µg/m³). While the CO 8-hour standard coincides with that of the USEPA, the NO_2 annual standard is more stringent than that of the USEPA (100 µg/m³) but lenient against that of the WHO (40 µg/m³). The 24-hour standard for SO_2 (70 µg/m³) is much more stringent than the EU limit value (125 µg/m³) and the USEPA standard

(370 µg/m^3) but lenient against the WHO guideline value (20 µg/m^3). The 24-hour PM$_{10}$ standard (120 µg/m^3) is lenient compared with the EU limit and WHO guideline value (50 µg/m^3) but more stringent than the US standard (150 µg/m^3). Nepal has not promulgated an annual standard for PM$_{10}$.

The 1993 Motor Vehicle and Transportation Management Act empowers local authorities to prescribe standards for the examination of vehicles and to fine those who violate traffic and transport rules. Since February 2000 the registration or transfer of ownership of 20-year-old vehicles in Nepal and the importation of reconditioned and second hand vehicles have been banned. The government has given 99 per cent tax exemption and full VAT exemption to owners of three-wheelers and 10–14 seater microbuses which run on less polluting alternative energy sources, such as CNG or liquid natural gas.

In 2000 Nepal adopted production emission standards for new private motor vehicles, light-duty vehicles, two-wheelers and three-wheelers. The standards define tests which include tests to verify exhaust emissions of CO, HC and NO$_x$ after a cold start, idling speed and the durability of pollution control devices (MOPE, 2005). Since October 2000 the 'Vehicle Emission Standards for Green Stickers' for in-use vehicles have been used to limit CO and HC in gasoline- and gas-operated two-, three- and four-wheelers. For diesel vehicles the Hartridge smoke unit has been limited.

In 2002, the Industrial Promotion Board of the Government decided that the government would stop registering brick kilns that use outdated Bulls Trench Kiln technology in the Kathmandu valley. The Board began the legal and administrative process to convert existing polluting industries to cleaner production options (CEN, 2002).

Emissions

There have been few studies on an inventory of air pollutants sources. A 1997 World Bank study estimated total emissions of TSP and PM$_{10}$ for 1993 and this was updated in 2002 for the base year 2001. The main sources of air pollutants in the Kathmandu Valley are motor vehicle exhausts, road re-suspension, brick manufacturing industries, domestic fuel combustion and refuse burning. There has been a decline in the production of bricks and the number of brick industries in operation. The total number of brick kilns in 1993 was 135 while in 2001 the number decreased to 116 (ENPHO, 2001). A major source of air pollution was the Himal Cement Factory, shut down in 2002 (KFW, 2004). While industrial sources of pollution have declined, other sources are increasing with the high rate of urbanization in the valley.

Monitoring

Kathmandu has six monitoring stations within the Kathmandu valley. As of 2004 only PM_{10} and NO_2 are being monitored routinely. The monitoring stations for PM_{10} automatically collect 24-hour samples through the eight filters mounted 3 metres above ground. The samples are collected once a week and analysed in a local laboratory (CEN/ENPHO, 2003). NO_2 is sampled by diffusion tubes exposed for a week. Other compounds (TSP, $PM_{2.5}$, SO_2, CO_2, benzene) are collected in campaigns and PAHs are analysed in campaigns as well (MOEST, 2005). A low-volume sampler without pneumatic movement designed according to the EN 12341 standard is used for PM_{10} and $PM_{2.5}$ sampling. It is equipped with a 24 volt pump in order to operate, with a battery back-up during power breaks. Power cuts are frequent in Nepal and can last from a few seconds to 12 hours (Gautam et al, 2004).

Measurements of NO_2 and SO_2 were undertaken by the Environmental Sector Programme Support (ESPS), a Nepal Denmark cooperation on industrial and urban environment, in February–March 2003 in the middle of the dry season, when the concentration is probably at its highest. The highest SO_2 level was recorded in Bhaktapur, where it was higher than 50 $\mu g/m^3$ at all times. In one of the four weeks the SO_2 concentration slightly exceeded the national 24-hour standard of 70 $\mu g/m^3$. At all the other stations the SO_2 levels remained below or around 50 $\mu g/m^3$ throughout the monitoring period (MOEST, 2005).

The results of the four-week monitoring showed that the NO_2 level was highest in Putali Sadak where it reached 50 $\mu g/m^3$ but at all the other stations the NO_2 levels were considerably lower than 50 $\mu g/m^3$. The high concentration of NO_2 at Putali Sadak is due to motor vehicle emissions. Routine measurements of NO_2 started in November 2003. Monthly averaged values in 2004 showed a peak of about 45 $\mu g/m^3$ in the dry season (December to May), and much lower concentrations during July to November 2004 (MOEST, 2005). These results indicate that annual NO_2 and SO_2 concentrations in the Kathmandu Valley could possibly comply with the national standards. Almost all studies undertaken so far in Kathmandu indicate that the levels of NO_2 and SO_2 are not of major concern (CEN/ENPHO, 2003).

Figure 3.10 shows the annual mean concentrations of PM_{10} at various sites in 2003 and 2004. The ambient concentrations of PM_{10} usually exceed the USEPA and EU annual standards and are five times above the WHO guideline value. Concentrations are significantly higher during the dry season (December–May 2004) compared to concentrations during the wet season (June–October 2004) (Gautam et al, 2004). Monthly PM_{10} concentrations exceed the national 24-hour standard at several sites throughout the monitoring period and would not comply with any monthly standard if it were set.

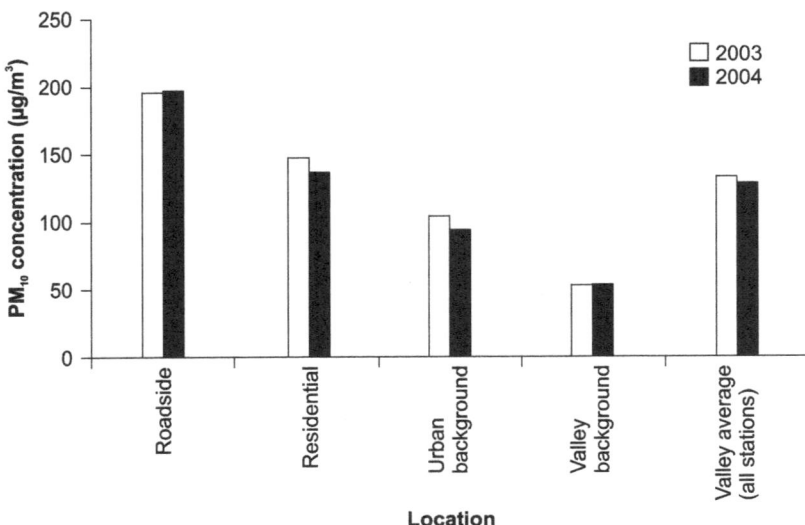

Figure 3.10 *Average annual concentrations of PM$_{10}$ in Kathmandu 2003–2004*

Source: MOEST (2005)

Health

The impact of air pollution on health in the Kathmandu valley has been assessed by examining the number of patients suffering from diseases related to air pollution. Medical records from hospitals in the Kathmandu valley revealed that urban residents have more respiratory diseases than rural residents (LEADERS, 1998). From 1996 to 1998, the number of patients with acute respiratory infection (ARI) increased by about 23 per cent per year. The share of ARI patients out of the total outpatient department visits was about 10 per cent in this period. Based on 1990 data, mortality in the Kathmandu valley due to air pollution is estimated to be 84 excess deaths annually while the number of respiratory symptom days was approximately 1.5 million. The annual cost of morbidity resulting from PM$_{10}$ is estimated to be US$6.1 million and total health damage to US$7.2 million (CEN, 2003).

A reduction of PM$_{2.5}$ levels in Kathmandu by half of the existing (47.4 µg/m^3) is estimated to result in a reduction in daily mortality of 7 per cent and hospital admissions of 24 per cent. Similarly, a reduction in the annual average PM$_{10}$ level in Kathmandu to the international standard (50 µg/m^3) will avoid over 2000 hospital admissions, over 40,000 emergency room visits, approximately 135,000 cases of acute bronchitis in children, over 4000 cases of chronic bronchitis and half a million asthma attacks.

The number of chronic obstructive pulmonary disease (COPD) patients admitted to three major hospitals in Kathmandu as well as the percentage of COPD patients as a percentage of total medical patients has increased significantly in the last ten years (CEN/ENPHO, 2003).

It is estimated that among the sources of air pollution, traffic (exhaust and re-suspension) has the largest impact on health. Furthermore, it is estimated that the reduction in vehicle exhaust emissions is effective in reducing health damage equivalent to US$11.61 per kg emission reduction (World Bank, 1997).

Approximately US$10 million in tourism revenue has been lost due to air pollution in the Kathmandu valley (World Bank, 1997; Business Age, 2001). The atmospheric data obtained from the Kathmandu airport from 1970 onwards show that there has been a substantial decrease in visibility in the valley since 1980 (Sapkota et al, 1997).

KOLKATA

UN estimated population 2003: 13.81 million
Area: 1785 km²
UN projected population 2015: 16.80 million
Climate: Tropical savannah
Map reference: 22° 32'N, 88° 21'E
Annual mean precipitation: 1582 mm
Altitude: 6 m
Annual mean temperature range: 13–36° C

SITUATIONAL ANALYSIS

General

Kolkata is the capital city of the state of West Bengal and the largest metropolitan city in the eastern part of India. It is situated in the Ganges Delta on the eastern bank of the Hooghly River, which divides the West Bengal region. The Kolkata Metropolitan Area (KMA) covers 1785 km², while the area covered by the core city, Kolkata Municipal Corporation, is only 187.33 km². There are 38 municipalities in the KMA and three municipal corporations, namely, Kolkata, Howrah and Chandannagar (CMDA, 2005). Its altitude above sea level is 5.8 metres. Kolkata has a tropical savannah climate.

Legislation

The 1981 Air (Prevention and Control of Pollution) Act amended in 1987 and the 1982 Air (Prevention and Control of Pollution) Rules provide the legal basis for air pollution abatement measures (MoEF, 2002, 2004a). The Air Act provides for the control and abatement of air pollution while the Air Rules define the procedures of the Central Pollution Control Board (CPCB) and State Pollution Control Boards (SPCBs). Additional key legislation and programmes relating to air pollution include: (i) The Environment (Protection) Act, 1986 (MoEF, 2004b); (ii) The National Policy Statement on Abatement of Pollution (1992); and (iii) The Environment Action Programme (1993) (Malé Declaration, 2000b).

National ambient air quality standards for SO_2, NO_2, CO, TSP, PM_{10} and lead with short-term (24-hours) and long-term (annual) limits have been set for industrial, residential (including rural and other areas) and sensitive areas in order of decreasing values for set limits (MoEF, 2004a). Except for the 24-hour standard for PM_{10} and the annual standard for SO_2, the Indian standards for industrial areas are more stringent than the USEPA standards, which refer to the general population. For CO 8-hour and 1-hour exposures the Indian standards in industrial areas and consequently in residential and sensitive areas are stricter than WHO guideline and EU limit values. The Indian annual standard for SO_2 in residential areas (60 µg/m³) is more lenient that the EU limit value, designed for the protection of ecosystems (20 µg/m³), however, this standard for sensitive areas (15 µg/m³) is more stringent. The 24-hour SO_2 standard in industrial areas (120 µg/m³), residential areas (80 µg/m³) and sensitive areas (30 µg/m³) is more stringent than the EU limit value (125 µg/m³) but more lenient than the WHO guideline value (20 µg/m³). In contrast, the annual mean for NO_2 in residential areas (60 µg/m³) is less stringent than the WHO guideline and EU limit value (40 µg/m³), but in sensitive areas it is more stringent (15 µg/m³). India does not have NO_2 standards for 1 hour but rather for an exposure time of 24 hours. For PM_{10}, Indian standards in the residential area (annual: 60 µg/m³; 24-hours: 100 µg/m³) are lenient compared to the EU limit values for exposure times of one year (40 µg/m³) and 24 hours (50 µg/m³) and the WHO guideline values (annual: 20 µg/m³; 24-hours 50 µg/m³). The annual PM_{10} standard for residential areas is more lenient than the USEPA standard (50 µg/m³) but the 24-hour standard is more stringent than the USEPA standard (150 µg/m³). The Indian annual lead standard in residential areas (0.75 µg/m³) is more lenient than the WHO guideline and EU limit value (0.5 µg/m³) but more stringent than the USEPA standard (1.5 µg/m³). No standard for O_3 has been adopted in India.

In 1999 the Kolkata High Court passed an order which required all private or commercial vehicles in KMA to comply with Bharat Stage II (Euro II equivalent) standard and all three- and two-wheelers to comply with India 2000 standard (Euro I equivalent) (WBPCB, 2003; CSE, 2006) In 2001 stricter emissions standards were introduced for new four-wheeled vehicles (with gross vehicle weight, GVW, 3500 kg) and all other vehicles (GVW > 3500 kg) in KMA with the exception of four-wheeled transport vehicles which have national, inter-state or tourist permits. Complementary to these emissions standards, auto-emission testing centres were equipped for improvement with the installation of a photo-imaging facility.

Since 2001 there has been a ban on the distribution and sale of loose 2T oil (inert mixture of mobil oil and petrol) in KMA to address the problem with two-stroke two- and three-wheelers. In addition, dedicated dispensers for the sale of pre-mixed 2T (low smoke) oil are now provided in Kolkata.

In 2001, under a notification by the West Bengal Pollution Control Board (WBPCB), the standards for PM emissions were set at 150 mg/Nm³ for all boilers irrespective of their steam generation capacity, for all ceramic kilns irrespective of the nature of kiln, and for all cast iron foundries (cupola furnaces), irrespective of their metal melting capacity, for all rolling mills (WBDoE, 2003).

Emissions

Industrial sources are responsible for approximately 48 per cent of the air pollution in Kolkata with mobile sources accounting for 50 per cent and the domestic sector 2 per cent (WBDoE, 2003). Large and medium industries emit approximately 56 per cent of PM, whereas the small units emit more than 40 per cent of PM consuming only 6 per cent of total coal used in the industrial sector in Kolkata every day (Chakraborti, 2003). Major sources of $PM_{2.5}$ are diesel and gasoline emissions, road dust re-suspension, coal and biomass burning, especially in the colder months (World Bank, 2004b).

Monitoring

Kolkata has 17 fixed active monitoring stations evenly distributed throughout the city which regularly monitor TSP, respirable particles (RSPM/considered an approximation of PM_{10}), SO_2, NO_x and lead and are operated mostly by the WBPCB (WBPCB, 2003). Three stations are operated and maintained by the National Environment Engineering Research Institute (NEERI). In 2003 the WBPCB installed five automatic ambient air quality monitoring units in Kolkata and in the industrial areas of Howrah, Haldia and Durgapur. Figure 3.11 presents the annual average concentrations for TSP, NO_2 and SO_2 for the period 1994–2004 and PM_{10} for 1997–2004. Since 1997 there has been a decreasing tendency in the annual mean concentrations of TSP and PM_{10} (RSPM); however, air quality standards for residential areas have been exceeded (140 µg/m³ and 60 µg/m³ respectively) (WBPCB, 2003; Bhattacharya, 2005). In contrast, NO_2 concentrations have been increasing and exceeded annual standards for residential areas (60 µg/m³) in 2001 and 2002 but complied with the standard in 2003 and 2004. Large and sudden variations in the NO_2 concentrations may indicate problems in data sampling. In the period 1998–2004 Kolkata achieved a significant reduction in annual SO_2 levels with average annual air quality standards for residential areas being met (60 µg/m³).

Since the nationwide ban on leaded gasoline in India in 2000, lead concentrations have declined substantially and meet the stringent air quality standard (0.75 µg/m³) (WBPCB, 2003).

Health

In 1995 an estimated 10,647 premature deaths were attributed to air pollution in Kolkata (Ghose, 2002). Studies have demonstrated that children inhaling polluted air in Kolkata suffer from adverse lung reactions and genetic abnormality in exposed lung tissues (Lahiri et al, 2000a; 2000b). Approximately 47 per cent of Kolkata's population

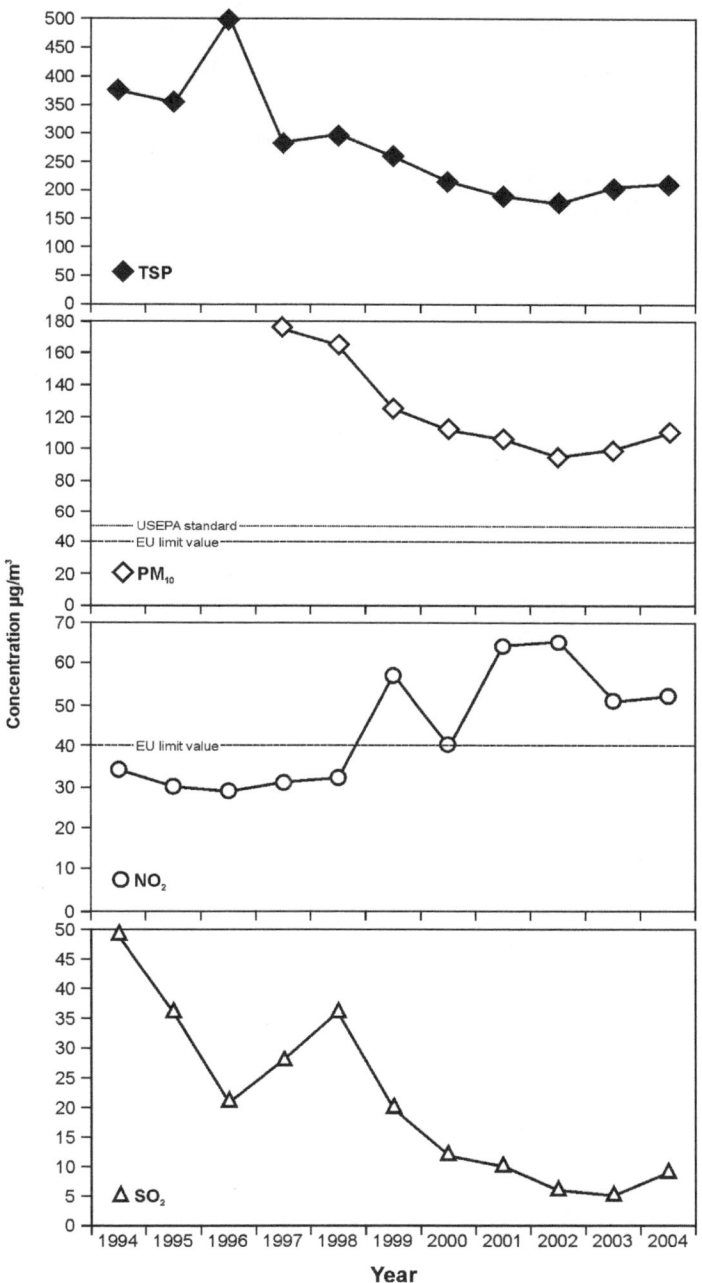

Figure 3.11 *Average annual concentrations of TSP, NO₂, SO₂, Kolkata 1992–2004, and PM₁₀, 1997–2004*

Source: WBPCB (2003); Bhattacharya (2005)

suffers from lower respiratory tract symptoms with the lungs of city residents being approximately seven times more burdened compared to their rural counterparts due to air pollution (Roy et al, 2001; WBPCB, 2003). Kolkata's street hawkers are at a higher risk of developing cancer due to the inhalation of benzo(a)pyrene over a period of eight hours each day sitting near strategic road junctions for an exposure time of 15 years (Chakraborti, 2003).

METRO MANILA

UN estimated population 2003: 10.35 million
UN projected population 2015: 12.64 million
Area: 636 km²
Climate: Tropical rain forest
Map reference: 14° 36′N 120° 59′E
Annual mean precipitation: 2159 mm
Altitude: 5 m
Annual mean temperature range: 21–34° C

SITUATIONAL ANALYSIS

General

Metropolitan Manila is the capital of the Philippines and is located along the shores of Manila Bay on the west side of Luzon Island. The metropolis is officially called the National Capital Region consisting of 14 cities and three municipalities covering an area of approximately 636 km². Metro Manila has a tropical rain forest climate and is approximately 5 metres above sea level.

Legislation

The 1999 Clean Air Act (CAA) provides the legal framework for air quality management in the country and established the air quality standards (CAA, 1999). The CAA mandates the Department of Environment and Natural Resources (DENR) through the Environmental Management Bureau as the department responsible for air pollution management, control and prevention. The DENR also collaborates with local governments under the 1991 Local Government Code (LGC, 1991).

The national air quality standards for PM_{10} (24-hours: 150 µg/m³; 1-year: 60 µg/m³) are lenient compared to the EU limit (24-hours: 50 µg/m³; 1-year: 40 µg/m³) and the WHO guideline values (24-hours: 50 µg/m³; 1-year: 20 µg/m³) but only the annual standard is lenient compared to the USEPA standard (50 µg/m³) while the 24-hour standard equals that of the USEPA. The SO_2 standard for 24 hours (180 µg/m³) is lenient compared to the EU limit value (125 µg/m³) and the WHO guideline value (20 µg/m³) but more stringent than the USEPA standard (365 µg/m³). The annual SO_2 standard (80 µg/m³) is also lenient compared to the EU limit value (20 µg/m³) for protection of the environment but equals the USEPA standard. There is no annual standard for NO_2 in the Philippines and the 24-hour standard has no equivalent in the USA, EU or WHO. The 8-hour O_3 standard of 60 µg/m^3 is more stringent than the WHO (100 µg/m^3) and EU (120 µg/m^3) values and the 1-hour O_3 standard (140 µg/m^3) is more stringent than the USEPA standard (240 µg/m^3). The CO standards are equivalent to the EU, WHO and USEPA limiting values. The Philippines have also adopted extreme short-term (30 to 60 minutes) ambient air quality standards for air pollutants emitted from industrial sources and operations such as ammonia, carbon disulphide, hydrogen chloride, hydrogen sulphide, SO_2, NO_2 and TSP. For SO_2 and NO_2 these values correspond roughly to those of the EU (CAA, 1999).

In 2003 stricter exhaust emissions limits, equivalent to Euro I standards, were introduced for new vehicles and motorcycles. Test procedures for these emissions are all based on Euro regulations (CAA, 1999; Walsh, 2000). There are plans to adopt Euro IV standard by 2009 (Walsh, 2000).

As part of the CAA, the annual inspection of vehicles is now required prior to registration. The Department of Transportation and Communications (DOTC) and Land Transportation Office (LTO) have been mandated to establish a Motor Vehicle Inspection System (MVIS); however, the establishment of such a system has yet to be implemented. By 2005, five Private Emission Testing Centers (PETCs) were created in order to conduct tests prior to registration, voluntary tests and tests of apprehended vehicles. These PETCs will operate until the establishment of a comprehensive MVIS (LTO, 2005).

Aside from the emissions testing that the vehicles undergo prior to registration, the vehicles may also be stopped on the road and subjected to smoke emission testing. In-use emissions standards are provided for gasoline- and diesel-fuelled vehicles. CO and HC are monitored for gasoline vehicles, while smoke is measured for diesel vehicles. In the period 2001–2003 there was a noticeable increase in the number of vehicles which passed the test (Santiago, 2003).

Emissions

In 2003 mobile sources in Metro Manila were responsible for 85 per cent of PM emissions while stationary sources were responsible for 15 per cent (Anglo, 2004). The food products industry is the largest contributor of industrial PM in Metro Manila, followed by textile companies (World Bank, 2002).

Monitoring

Metro Manila has had a network of nine stations monitoring air quality since 2004. These stations, funded by ADB, measure real-time PM_{10}, SO_2, NO_2, O_3, CO, benzene, xylene, toluene, methane, non-methane hydrocarbons and total hydrocarbons as well as meteorological conditions (DENR, 2005).

Concentrations of SO_2 have been significantly reduced in Metro Manila and currently comply with the air quality standard (80 µg/m³) (Amador, 2003; DENR, 2005). In the period 2000–2004 $PM_{2.5}$ concentrations monitored in various locations within Metro Manila (residential, traffic, commercial, industrial, agricultural and mixed land use areas) and averaged over this period showed that at traffic and industrial locations $PM_{2.5}$ concentrations did not comply with the USEPA 24-hour (65 µg/m³) and annual standards (15 µg/m³, respectively) (Villarin et al, 2004). Figure 3.12 shows the annual concentrations of TSP for the period 1994–2004.

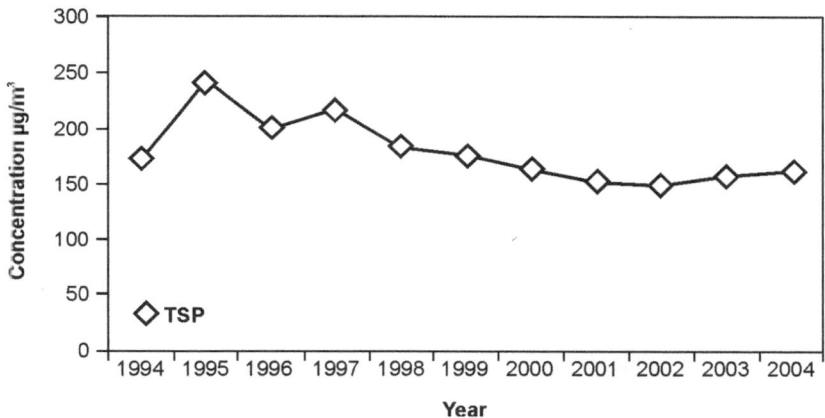

Figure 3.12 *Average annual concentrations of TSP in Metro Manila 1994–2004*

Source: Villarin et al (2004), DENR (2005)

Annual mean concentrations of TSP exceed the national guideline value for TSP (90 µg/m³) (DENR, 2005; Vergel, 2004). Annual averages of PM_{10} concentrations monitored in 2002 and 2003 varied between 50 and 60 µg/m³. Ambient monitoring of NO_x as well as CO, O_3 and lead is not well established in Metro Manila (DENR, 2005).

Health

Children exposed to high levels of TSP (259 µg/m³ per 24 hours) have developed a history of allergy and nasal congestion, while at low/fair levels of TSP (130 µg/m³) sneezing was observed (Briones, 2002). An estimated additional 10,000 cases of acute bronchitis, 300 cases of asthma, 9 cases of chronic bronchitis, 200 respiratory hospital admissions, 40 cardiovascular admissions, up to 200 cardiovascular deaths and up to 330 respiratory deaths were attributed to PM_{10} levels (Torres et al, 2004) annually. In the Philippines the impacts and costs on health attributed to air pollution, particularly PM_{10}, were estimated at US$392 million for 2001, based on the monetary costs for excess deaths (1915), treatment of chronic bronchitis incidences (8439) and respiratory symptom incidences (50.5 million) linked to PM_{10} (Torres et al, 2004).

MUMBAI

UN estimated population 2003: 17.43 million
UN projected population 2015: 22.65 million
Area: 437.71 km²
Climate: Tropical savannah
Map reference: 18° 56'N, 72° 51'E
Annual mean precipitation: 2078 mm
Altitude: 11 m
Annual mean temperature range: 7–41° C

SITUATIONAL ANALYSIS

General

Mumbai is located on the west coast of India and consists of a peninsula originally composed of seven islets. Mumbai is approximately 11 metres above sea level and consists of several islands on the Konkan coast. The city (Greater Mumbai) occupies an area of approximately 437.71 km². Mumbai has a tropical savannah climate. High pollution concentrations usually occur in the winter due to adverse meteorological situations and low wind speed (MCGM, 2004).

Legislation

The 1981 Air (Prevention and Control of Pollution) Act amended in 1987 and the 1982 Air (Prevention and Control of Pollution) Rules provide the legal basis for air pollution abatement measures (MoEF, 2002; 2004a). The Air Act provides for the control and abatement of air pollution while the Air Rules define the procedures of the Central Pollution Control Board (CPCB) and State Pollution Control Boards (SPCBs). Additional key legislation and programmes relating to air pollution include: (i) The Environment (Protection) Act, 1986 (MoEF, 2004b); (ii) The National Policy Statement on Abatement of Pollution (1992); and (iii) The Environment Action Programme (1993) (Malé Declaration, 2000b).

National ambient air quality standards for SO_2, NO_2, CO, TSP, PM_{10} and lead with short-term (24-hours) and long-term (annual) limits have been set for industrial, residential (including rural and other areas) and sensitive areas in order of decreasing values for set limits (MoEF, 2004a). Except for the 24-hour standard for PM_{10} and the annual standard for SO_2, the Indian standards for industrial areas are more stringent than the USEPA standards, which refer to the general population. For CO 8-hour and 1-hour exposures the Indian standards in industrial areas, and consequently in residential and sensitive areas, are stricter than WHO guideline and EU limit values. The Indian annual standard for SO_2 in residential areas (60 $\mu g/m^3$) is more lenient that the EU limit value, designed for the protection of ecosystems (20 $\mu g/m^3$), however, this standard for sensitive areas (15 $\mu g/m^3$) is more stringent. The 24-hour SO_2 standard in industrial areas (120 $\mu g/m^3$), residential areas (80 $\mu g/m^3$) and sensitive areas (30 $\mu g/m^3$) is more stringent than the EU limit value (125 $\mu g/m^3$) but more lenient than the WHO guideline value (20 $\mu g/m^3$). In contrast, the annual mean for NO_2 in residential areas (60 $\mu g/m^3$) is less stringent than the WHO guideline and EU limit value (40 $\mu g/m^3$), but in sensitive areas it is more stringent (15 $\mu g/m^3$). India does not have NO_2 standards for 1 hour but rather for an exposure time of 24 hours. For PM_{10}, Indian standards in the residential area (annual: 60 $\mu g/m^3$; 24-hours: 100 $\mu g/m^3$) are more lenient than the EU limit values for exposure times of one year (40 $\mu g/m^3$) and 24 hours (50 $\mu g/m^3$) and the WHO guideline values (annual: 20 $\mu g/m^3$; 24-hour: 50 $\mu g/m^3$). The annual PM_{10} standard for residential areas is more lenient than the USEPA standard (50 $\mu g/m^3$) but more stringent for 24 hours than the USEPA standard (150 $\mu g/m^3$). The Indian annual lead standard in residential areas (0.75 $\mu g/m^3$) is more lenient than the WHO guideline and EU limit value (0.5 $\mu g/m^3$) but more stringent than the USEPA standard (1.5 $\mu g/m^3$). No standard for O_3 has been adopted in India.

The Air and Environment Acts stipulate that the setting of automobile emissions standards is the responsibility of the CPCB or Ministry of Environment and Forests. The Transport Commissioner is responsible for the implementation and enforcement of these standards under the 1988 Motor Vehicles Act and 1989 Central Motor Vehicles Rules.

Mass emission standards for new vehicles were first introduced in India in 1991. Stringent emission norms along with fuel quality specifications were laid down in 1996 and 2000. Euro I vehicle emission standards were applicable from 1 April 2000 and Euro II standards were applicable all over India from 1 April 2005. In Mumbai Euro II standards for all new non-commercial vehicles were adopted in October 2002 for all commercial vehicles (World Bank, 2004a).

Lead in gasoline was completely phased out in Mumbai and the rest of India in February 2000. In 2000, the benzene and sulphur content in gasoline was reduced to 1 per cent and 0.25 per cent, respectively. Mumbai reduced sulphur in diesel to 0.25 per cent in 1998, and to 0.05 per cent in April 2000 (Ghose, 2002; Singh, 2002).

The Mumbai Pollution Control Board (MPCB) is responsible for implementing the industry-specific discharge and emission standards, also commonly referred to as MINAS (Minimum National Standards), as prescribed by the CPCB. Other institutions and policies related to stationary sources which have been implemented in Mumbai include the Environmental Safety Committee, established after the Bhopal accident, which provides experts for safety inspection of major plants. In addition, the 1984 Industrial Location Policy of Mumbai Metropolitan Region prohibits the expansion of large-, medium- and small-scale units in Mumbai and the 1979 Use of Coal Rule of the Urban Development Department prohibits the issuing of new permits for coal use in Mumbai (World Bank, 1997).

Emissions

A comprehensive emissions inventory for Mumbai was developed under the URBAIR project (World Bank, 1997). Major sources of TSP were identified as being re-suspension of road dust (approximately 40 per cent), wood combustion (17 per cent), domestic refuse burning (approximately 14 per cent) and diesel vehicle exhaust (9 per cent). The main source of SO_2 emissions was industrial fuel oil combustion and power plants (82 per cent) while motor vehicles and industrial fuel combustion were major sources of NO_x emissions (52 and 41 per cent, respectively). More recently, stone crushers have been identified as accounting for 37 per cent of TSP emissions (NEERI, 2005). Receptor modelling was conducted to provide reasonable order-of-magnitude estimates of the contribution of different sources to RSPM (PM_{10}) levels (World Bank, 2004a).

Monitoring

The Municipal Corporation of Greater Mumbai operates and maintains 22 monitoring stations in the city. These were manually operated on a non-continuous basis to monitor the criteria air pollutants, namely SO_2, NO_2, TSP and ammonia. In addition, NEERI has operated three monitoring stations (World Bank, 1997). Figure 3.13 presents the annual averages of TSP, PM_{10}, SO_2 and NO_2 for 1994–2004.

Annual levels of TSP fluctuate around 250 µg/m³ and do not comply with the Indian standard of 140 µg/m³. PM_{10} (RSPM) levels in Mumbai have continued to exceed the national standard for annual averages in residential areas (60 µg/m³). Between 1997 and 2002 PM_{10} levels had a decreasing tendency with the ambient concentrations for commercial, industrial and residential areas nearly approaching the annual ambient standard for residential areas. After 2002, concentrations of PM_{10} levelled off and show a slight increase.

Between 1998 and 2004 the annual average concentrations of NO_2 in Mumbai declined and they are well below the India standard (80 µg/m³). Between 1997 and 2004 SO_2 concentrations also declined. The annual mean levels are very low and comply with the standard (80 µg/m³).

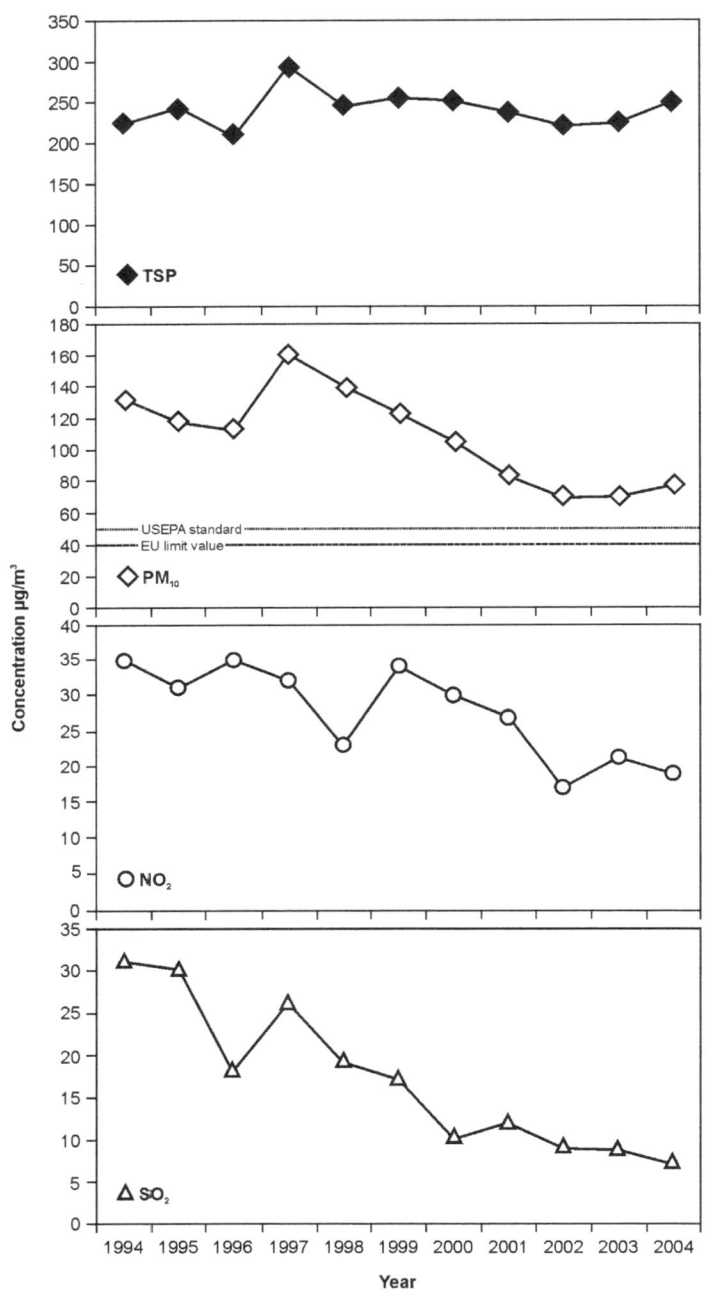

Figure 3.13 *Average annual concentrations of TSP, PM$_{10}$, NO$_2$, SO$_2$, Mumbai 1994–2004*

Source: NEERI (2005)

Health

Several epidemiological studies have shown that in three urban communities in Mumbai, with moderately raised pollution levels (SO_2, NO_2, TSP), there was an increased occurrence of dyspnoea, chronic and intermittent cough, frequent colds, chronic bronchitis and cardiac disorders, mainly cough, high blood pressure and ischaemic heart disease, and deaths due to non-tuberculous respiratory and ischaemic heart disease (Kamat, 2000). Significant health differentials were also shown to exist between a high-polluted area and a low or less polluted area in Mumbai (Shankar and Rao, 2002).

The cost of health impacts in Mumbai have been estimated to be a total of US$44.9 million which is attributed to mortality using the 'loss in salary approach' and US$206.3 million for costs attributed to morbidity (Srivastava and Kumar, 2002). Another study has shown that the pollution in Mumbai can lead to high health costs of approximately 19 per cent of the income of patients suffering from severe air pollution-related attacks (Parikh and Hadker, 2003).

NEW DELHI

UN estimated population 2003: 14.15 million
UN projected population 2015: 20.95 million
Area: 1483 km²
Climate: Tropical steppe
Map reference: 28° 38'N, 77° 17'E
Annual mean precipitation: 60 mm
Altitude: 216 m
Annual mean temperature range: 7–41° C

SITUATIONAL ANALYSIS

General

New Delhi is the capital city of India. It lies along the banks of river Yamuna in the northern part of India south of the Himalayas. It covers a total area of 1483 km² and is approximately 216 metres above sea level. New Delhi has a tropical steppe climate. Due to westerly winds, it is affected by dust storms, which leads to hazy atmosphere and high ambient levels of PM (CPCB, 1998).

Legislation

The 1981 Air (Prevention and Control of Pollution) Act amended in 1987 and the 1982 Air (Prevention and Control of Pollution) Rules provide the legal basis for air pollution abatement measures (MoEF, 1981a; 1981b; 1983). The Air Act provides for the control and abatement of air pollution while the Air Rules define the procedures of the Central Pollution Control Board (CPCB) and State Pollution Control Boards (SPCBs). Additional key legislation and programmes relating to air pollution include: (i) The Environment (Protection) Act, 1986 (MoEF, 2004); (ii) The National Policy Statement on Abatement of Pollution (1992); and (iii) The Environment Action Programme (1993) (Malé Declaration, 2000a). In 1997, an action plan for cleaner air in New Delhi was developed (MoEF, 1997).

National ambient air quality standards for SO_2, NO_2, CO, TSP, PM_{10} and lead with short-term (24 hours) and long-term (annual) limits have been set for industrial, residential (including rural and other areas) and sensitive areas in order of decreasing values for set limits (MoEF, 2004b). Except for the 24-hour standard for PM_{10} and the annual standard for SO_2, the Indian standards for industrial areas are more stringent than the USEPA standards, which refer to the general population. For CO 8-hour and 1-hour exposures the Indian standards in industrial areas and consequently in residential and sensitive areas are stricter than WHO guidelines and EU limit values. The Indian annual standard for SO_2 in residential areas (60 $\mu g/m^3$) is more lenient that the EU limit value, designed for the protection of ecosystems (20 $\mu g/m^3$); however, this standard for sensitive areas (15 $\mu g/m^3$) is more stringent. The 24-hour SO_2 standard in industrial areas (120 $\mu g/m^3$), residential areas (80 $\mu g/m^3$) and sensitive areas (30 $\mu g/m^3$) is more stringent than the EU limit value (125 $\mu g/m^3$) but more lenient than the WHO guideline value (20 $\mu g/m^3$). In contrast, the annual mean for NO_2 in residential areas (60 $\mu g/m^3$) is less stringent than the WHO guideline and EU limit value (40 $\mu g/m^3$), but in sensitive areas it is more stringent (15 $\mu g/m^3$). India does not have NO_2 standards for 1 hour but rather for an exposure time of 24 hours. For PM_{10}, Indian standards in the residential area (annual: 60 $\mu g/m^3$; 24-hours: 100 $\mu g/m^3$) are more lenient than the EU limit values for exposure times of one year (40 $\mu g/m^3$) and 24 hours (50 $\mu g/m^3$) and the WHO guideline values (annual: 20 $\mu g/m^3$; 24-hour: 50 $\mu g/m^3$). The annual PM_{10} standard for residential areas is more lenient than the USEPA standard (50 $\mu g/m^3$) but more stringent for 24 hours than the USEPA standard (150 $\mu g/m^3$). The Indian annual lead standard in residential areas (0.75 $\mu g/m^3$) is more lenient than the WHO guideline and EU limit value (0.5 $\mu g/m^3$) but more stringent than the USEPA standard (1.5 $\mu g/m^3$). No standard for O_3 has been adopted in India.

Indian vehicle emissions standards for four-wheelers are based on European standards. It has its own standards for two- and three- wheelers (Mashelkar et al, 2002). Since 2001 Bharat Stage II (equivalent to Euro II) emissions standards for new vehicles have been in place in New Delhi. In April 2005 these standards were implemented for the entire country. In addition Bharat Stage III and Bharat Stage II standards for two- and three-wheelers have been introduced (FADA, 2005).

Vehicle emission standards for new vehicles were first introduced in 1991 and subsequently revised in 1996 and 2000. In October 2004 revised idle emission standards for on-road vehicles come into effect.

In 2001 the Ministry of Road Transport and Highways (MRTH) issued revised emission standards for new vehicles using compressed natural gas (CNG) and in-use vehicles converted to CNG. The MRTH has also introduced emissions standards for liquid petroleum gas (LPG) (Dursbeck et al, 2001).

In 2000 gasoline and diesel improvements were implemented in New Delhi. Sulphur in gasoline was reduced from 0.10 per cent to 0.05 per cent and benzene levels from 3 per cent to 1 per cent. Sulphur in diesel was reduced from 0.25 per cent to 0.05 per cent (Mashelkar et al, 2002).

Fuels used by industry allowed in metropolitan New Delhi are low-sulphur (0.4 per cent) coal, fuel oil/light diesel oil (LDO) and low sulphur heavy stock (LSHS) with low sulphur (1.8 per cent). Other fuels used are LPG, CNG, kerosene, naphtha (a light gasoline) for power stations, aviation engine for aircrafts, firewood for domestic use in rural areas and crematoriums and biogas (CPCB, 2005b).

Emissions

In New Delhi thermal plants are responsible for approximately 68 per cent of SO_2 emissions and 80 per cent of TSP emissions. The transport sector is a main contributor to NO_x, CO (70–90 per cent) and non-methane VOCs (80 per cent) (Mashelkar et al, 2002; Gurjar et al, 2003). A receptor modelling study of PM_{10} indicated that there were three principal sources of particle air pollution: vehicle exhaust, re-suspended road dust and solid fuels (World Bank, 2004b).

Monitoring

The CPCB and NEERI share the operation and maintenance of the existing 11 monitoring stations (CPCB 2001a; 2001b; 2003a; 2003b). All compounds, with the exception of NO_2, show a decreasing tendency. Roadside CO levels approach the national air quality standard (2000 µg/m³) for residential 8-hour exposure to CO (GONCT, 2003). Ambient lead levels also decreased as a consequence of lead phase-out in gasoline. Figure 3.14 shows the annual average concentrations of TSP, PM_{10}, NO_2 and SO_2 for the period 1994–2004 (CPCB, 2005b).

TSP levels fluctuate between 300 and 360 µg/m³, well above the annual standard of 140 µg/m³. Annual PM_{10} concentrations decreased from 270 µg/m³ in 1994 to half this level in 2004, some 15 per cent above the Indian standard of 120 µg/m³. PM_{10} levels, which have been recently monitored, also are double that of the national 60 µg/m³ annual limit for residential areas (Duggal and Pandey, 2002; World Bank, 2004a).

Since 2001 annual NO_2 concentrations have exhibited a slightly increasing tendency with values well below the national standard (60 µg/m³) for residential areas (World Bank, 2004a). The annual SO_2 concentrations decreased steadily from 1994 to 2004 and are relatively low and comply with the air quality standard (60 µg/m³) for residential areas.

Health

The prevalence of chronic respiratory symptoms and diseases in the city is substantially greater in those individuals living in slums and low-income housing areas. One study showed that an increase in the economic status of the individual significantly decreased the prevalence of respiratory symptoms among those who were exposed long term to high levels of air pollution (Chhabra et al, 2001).

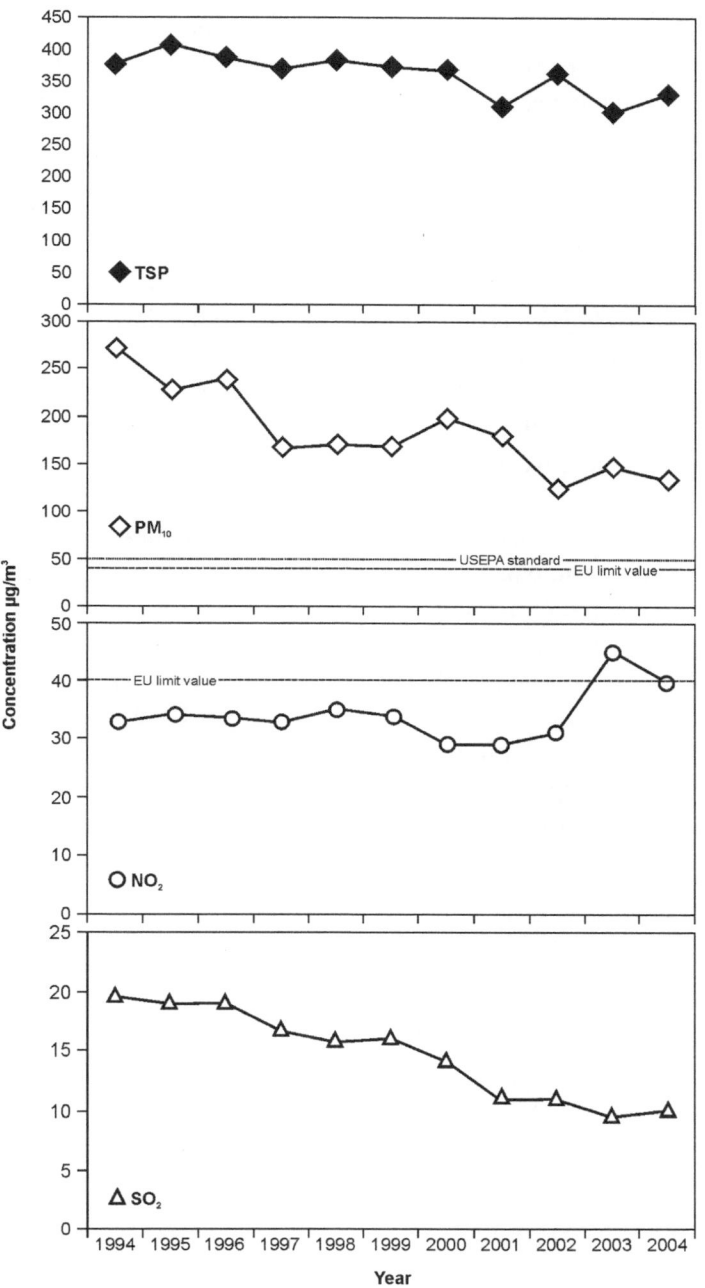

Figure 3.14 *Average annual concentrations of TSP, PM$_{10}$, NO$_2$, SO$_2$,*
New Delhi 1994–2004

Source: CPCB (2005c)

A time-series study of the impact of particulate air pollution on daily mortality in New Delhi showed a positive significant relationship between particulate pollution and daily non-traumatic deaths, as well as deaths from respiratory and cardiovascular problems and for certain age groups (Cropper et al, 1997).

A study conducted by the All India Institute of Medical Sciences in New Delhi showed that exposure to PM has an impact on respiratory health and contributes to respiratory morbidity. Results indicated that most common symptoms related to air pollution were irritation of eyes, cough, pharyngitis, dyspnoea and nausea. The increase in hospital admissions due to respiratory morbidity have also been correlated with a rise in PM levels (Mashelkar et al, 2002).

Agarwal et al (2002) compared the vitamin D status of infants and toddlers in two areas in New Delhi with a high and a low level of air pollution. The mean serum concentration of 25-hydroxyvitamin D of infants in the lower pollution area was more than double that of children living in the highly polluted area, suggesting that air pollution may be the cause of the lower serum concentration of vitamin D in the highly exposed infant sample.

The National Institute of Public Finance and Policy was commissioned by the Expert Committee on Auto Fuel Policy to investigate the costs of health damages from vehicle emissions. The results showed that the annual health damage of pre-Euro emissions for 25 Indian cities ranged from US$14 million (Rs 6.8 billion) to US$191.6 million (Rs 93.1 billion) (Mashelkar et al, 2002). The government has shown the commitment to further investigate the impacts of air pollution by initiating a comparative risk assessment study, together with WHO, of nine Indian cities.

Seoul

SOUTH KOREA

SEOUL

UN estimated population 2003: 9.71 million
UN projected population 2015: 9.21 million
Area: 605.52 km²
Climate: Continental warm summer
Map reference: 37° 32'N, 127° 00'E
Annual mean precipitation: 1200 mm
Altitude: 60 m
Annual mean temperature range: –7–31° C

SITUATIONAL ANALYSIS

General

Seoul is the capital city of the Republic of Korea. It is located 30 km inland of the Yellow Sea (Huang Hai) on the north side of the river Han (Hangang) and is less than 100 metres above sea level. The total area of Seoul is 605.52 km². The Hangang River bisects the city in two parts: the northern part (Gangbok) and the southern part (Gangnam). The Gangbok region totals 297.97 km² while Gangnam is 307.55 km². Seoul has a continental warm climate.

Legislation

The 1977 Environment Conservation Law provided the basis for the 1990 Clean Air Conservation Law (CACL) which requires mandatory continuous monitoring of air pollutants and stipulates emission limits and the procedures for the authorization of air pollutant emissions from different sources. Under this law, permission must be obtained from the regional environmental office for major private and public developments or smaller scale projects that are to be located in areas which do not meet national air quality standards. An environmental impact assessment must be conducted to ensure environmental considerations are to be included in the development plans so as to minimize any potential impacts (MoL, 2002).

National air quality standards exist for SO_2, CO, NO_x, O_3, Pb and PM. However, due to differences in regional characteristics, local governments of each province have the authority to enforce their own municipal ordinances for environmental quality standards. In March 1998 Seoul implemented more stringent air quality standards for SO_2, NO_x and PM_{10} (MoL, 2002). The Seoul annual air quality standards for NO_x (75 µg/m³) and lead (0.5 µg/m³) are more stringent than the USEPA standards (100 µg/m³ [for NO_2] and 1.5 µg/m³, respectively). The NO_x annual standard is less stringent than the EU limit value of 40 µg/m³ for NO_2. The annual air quality standard for SO_2 (27 µg/m³) is also more stringent than the USEPA standard (80 µg/m³) and almost as stringent as the EU limit value (20 µg/m³), which, however, is designed for the protection of ecological systems. The 24-hour SO_2 standard (105 µg/m³) is more stringent than both the USEPA standard (370 µg/m³) and EU limit value (125 µg/m³) but is more lenient than the WHO guideline value (20 µg/m³). The Korean 8-hour standard for CO (10,170 µg/m³) and O_3 (120 µg/m³) are equal or approximately equal to those of the EU; the O_3 1-hour standard (200 µg/m³) is slightly more stringent than the USEPA standard (240 µg/m³). The Seoul annual PM_{10} standard of 60 µg/m³ is less stringent than the corresponding USEPA standard (50 µg/m³), EU limit value (40 µg/m³) and WHO guideline value (20 µg/m³). The Korean 24-hour standard for PM_{10} (120 µg/m³) is more stringent than that of the USEPA (150 µg/m³) but is more than double that of the EU limit and WHO guideline value (50 µg/m³).

The current emission standards for new vehicles powered by gasoline and diesel fuel meet Euro III and 2002 TLEV standards and will be strengthened to Euro IV and ULEV standards by 2006. From July 2002 the PM emission standard was lowered to between 0.20 g/kWh and 0.10 g/kWh for heavy-duty diesel vehicles. From 2006 the standards will be lowered further to 0.02 g/kWh. Vehicles undergo inspections at least every two years and are subject to mandatory maintenance orders with the possibility of fines if any of the pollutants are exceeded. Random roadside checks are also conducted with over 200 inspection teams based around Korea. The sulphur content of diesel fuel was lowered to 0.043 per cent (430 ppm) in 2002 and will be further lowered to 30 ppm by 2006.

For stationary sources, the government has implemented policies to reduce the sulphur content in fuels used by industry in order to meet the objective of lowering the annual mean SO_2 concentration below the standard of 55 µg/m³ (20 ppb). To manage air pollutants emitted from manufacturing facilities, emission standards for 18 types of gaseous substances, 9 types of particulate substances and 8 noxious substance types equivalent to EU and Japanese standards were established out of the total of 52 pollutants specified in the CACL (MOE, 2006).

Emissions

In Seoul, mobile sources account for approximately 89.3 per cent of CO, 59.0 per cent of NO_x and 71.9 per cent of PM_{10} emissions. Sources for SO_2 include non-industrial combustion (67.3 per cent), combustion from manufacturing industry (14.6 per cent), on-road mobile sources (8.0 per cent) and district-wide heating plants (6.1 per cent) (MOE, 2003).

Monitoring

In 2002 Seoul had 27 monitoring stations which continuously measured six air pollutants including PM_{10}, CO, O_3, SO_2, NO_x and dust (APMA/KEI, 2002). At 11 stations, VOCs detectors were installed to measure the amount of ozone-forming pollutants (MOE, 2003).

Since the 1990s SO_2 and lead levels decreased dramatically and complied with Seoul's air quality standards. The annual levels for PM_{10}, O_3, NO_2 and CO are exhibited in Figure 3.15 (APMA/KEI, 2002; MOE, 2002; 2003).

Despite stringent emission standards, PM_{10} levels are stagnant and do not yet comply with Seoul's standard of 60 $\mu g/m^3$. This is due to the increasing number of motor vehicles and the impact of the yellow sand from dust and sandstorms from mainland China. NO_2 annual levels have shown an increasing tendency. Annual means for CO decreased substantially between 1994 and 2004, having reached values below 800 $\mu g/m^3$.

Health

In spring, transboundary air pollution combined with dust storms from the deserts of northern China and Mongolia, has resulted in serious health impacts in Seoul. Research tracing the rate of deaths among Seoul residents from March to May, 1995–1998, showed the death rate on 'yellow sand days' to be 1.7 per cent greater than on normal days, everything else being equal. During the study period, PM_{10} concentrations on 'yellow sand days' averaged 101 $\mu g/m^3$ compared to an average of 73 $\mu g/m^3$ on non-yellow sand days (Kwon et al, 2002).

Increases in daily total mortality due to ambient air pollution have been observed in Seoul (Kwon and Cho, 1999). Kwon et al (2002) estimated the effect of air pollution on daily mortality of patients with congestive heart failure among residents of Seoul during the period 1994–1998 in comparison with that of the general population in the same area and the same period. The estimated effects appeared larger among the congestive heart failure patients than among the general population (2.5 to 4.1 times higher depending on the pollutant considered).

In Seoul, SO_2 has been identified as a significant predictor for all-cause deaths. An increase of 133 $\mu g/m^3$ (50 ppb) of SO_2 corresponded to a 3.9 per cent increase in excess deaths (95 per cent confidence level: 0.7–7.2 per cent). This relationship does not change if weather conditions and the other pollutants are included in the regression (Lee et al, 2000). Assuming that the mortality in Seoul equals that of Korea (520/100,000), approximately 1967 excess deaths per year (95 per cent confidence interval: 353–3635) would be associated with an increase of 133 $\mu g/m^3$ in SO_2. This finding is consistent with those of similar studies in Seoul (Lee et al, 1999; Lee and Schwartz, 1999).

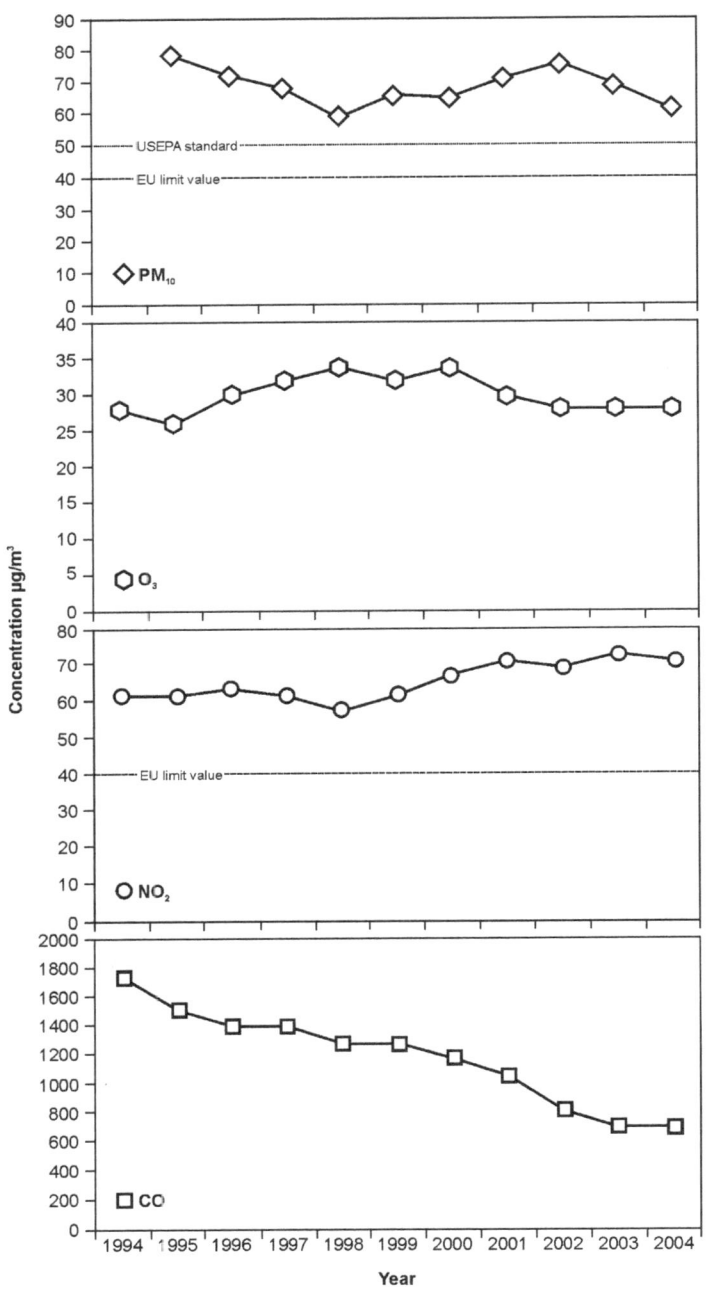

Figure 3.15 *Average annual concentrations of PM$_{10}$, O$_3$, NO$_2$, CO, Seoul 1994–2004*

Source: MOE (2006)

In Seoul the average social costs per kg of emitted compounds are approximately US$25.6 per kg of PM_{10}, US$8.8 per kg of SO_2, US$7.9 per kg NO_x, US$6.5 per kg CO and US$7.6 per /kg VOCs, respectively (MOE, 2002). When these estimates are applied to Seoul's case, the social costs caused by overall air pollution are approximately US$2.5 billion, with CO comprising the largest portion of the costs at 40.5 per cent.

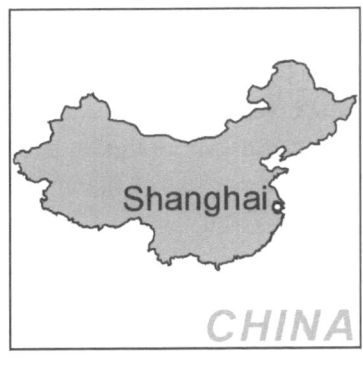

SHANGHAI

UN estimated population 2003: 12.76 million
UN projected population 2015: 12.67 million
Area: 6340 km²
Climate: Humid sub-tropical
Map reference: 31° 14'N, 121° 30'E
Annual mean precipitation: 1300 mm
Altitude: 4.5 m
Annual mean temperature range: 1–32 °C

SITUATIONAL ANALYSIS

General

Shanghai is located on the Yangtze River Delta and is bordered by Jiangsu and Zhejiang provinces to the west and the East China Sea to the east. It is 4.5 metres above sea level and has a total area of approximately 6340 km², 11 per cent of which consists of rivers and lakes. Shanghai has a humid sub-tropical climate.

Legislation

The 1989 national Environmental Protection Law is the framework law for the prevention and control of pollution. However, the 1987 Law on Prevention and Control of Atmospheric Pollution sets out regulations for air pollution including environmental impact assessment (Article 11); total emission control and emission licensing (Article 15); prohibition of industrial sources near protected areas (Article 16); banning of equipment causing serious air pollution (Article 19); prohibition of coal washing (Article 24); application of emission standards and controls for fixed and mobile sources (Articles 27, 30, 32).

National ambient air quality standards are categorized into three classes. Designated industrial areas are expected to comply with Class III standards, residential areas with Class II standards and parks and specially protected areas with Class I. Regulation GB3095-1996 requires Shanghai to comply with Class II standards. The annual Class II standards for SO_2 (60 µg/m³) and NO_2 (40 µg/m³) are more stringent than those of the USEPA (80 µg/m³ for SO_2; 100 µg/m³ for NO_2), but are more lenient or equal to the EU limit values (20 µg/m³ for SO_2; 40 µg/m³ for NO_2). The EU limit for SO_2 protects ecosystems and corresponds to the Class I standard. The Class II standards for 24-hour exposure to SO_2 (150 µg/m³) are lenient compared with the WHO guideline (20 µg/m³) and EU limit values (125 µg/m³) but more stringent than the USEPA standard of 365 $\mu g/m^3$. The Class II 1-hour standard for SO_2 (500 $\mu g/m^3$) is lenient compared with the EU limit value of 350 $\mu g/m^3$ as is the 1-hour standard for NO_2 (240 $\mu g/m^3$) compared with the WHO guideline and EU limit values (200 $\mu g/m^3$). For O_3 the Chinese standard (160 $\mu g/m^3$) is more stringent than the USEPA standard (235 $\mu g/m^3$) but more lenient than the EU limit (120 $\mu g/m^3$) and the WHO guideline value (100 $\mu g/m^3$). The PM_{10} standards for 1 year and 24 hours (100 $\mu g/m^3$, 150 $\mu g/m^3$) are more lenient than the EU limit values (40 $\mu g/m^3$, 50 $\mu g/m^3$), the WHO guideline values (20 $\mu g/m^3$, 50 $\mu g/m^3$) and the USEPA annual standard (50 $\mu g/m^3$) but are equal to the USEPA 24-hour standard (150 $\mu g/m^3$).

Use of leaded gasoline in Shanghai has been banned since 1997 (Wang, 2003). Chinese vehicle emission regulation (GB-18352-1), equivalent to Euro I standards was enacted in April 2001. Stricter Euro II equivalent vehicle emission standards have also been imposed in Shanghai since 2003 (Jingguang, 2003). In 2005 all gasoline two-wheelers were phased out in Shanghai although a few are still operating in the suburban areas (ADB, 2003).

For stationary and area sources, emissions from thermal power plants; cement plants; coal-burning, oil-burning and gas-fired boilers; coke ovens; industrial kilns and furnaces and even cooking fumes are subject to emissions standards. Regulations and standards are also being implemented for solid waste – specifically for incineration of municipal solid wastes and hazardous wastes; storage and disposal sites of solid wastes from general industries and on pollutants from fly ash use in agriculture (SEPA, 2003).

Emissions

In 2001 industrial sources emitted a total volume of 472,600 tons of SO_2, 135,200 tons of particulate from coal burning and 18,200 tons of dust, which were already lower than the previous decade by 41,000 tons (for SO_2) and 94,000 tons (particulate from industry) (Wang, 2003). In Shanghai, power plants are major emitters of SO_2 and NO_x (approximately 53 per cent); other industries contributed 20–26 per cent. Industrial PM emissions amount to approximately 20 per cent of total emissions (Fu, 2004).

Monitoring

The Shanghai Environmental Monitoring Centre (SEMC) is responsible for a total of 44 monitoring stations in Shanghai, 21 of which are automatic. Pollutants monitored include TSP, PM_{10}, CO, SO_2, NO_x, lead, dust and fluoride (UNESCAP, 2000).

Figure 3.16 shows the ambient air pollutant concentrations for TSP, PM_{10}, NO_2 and SO_2. TSP levels decreased from 281 $\mu g/m^3$ in 1994 to 140 $\mu g/m^3$ in 2003. Between

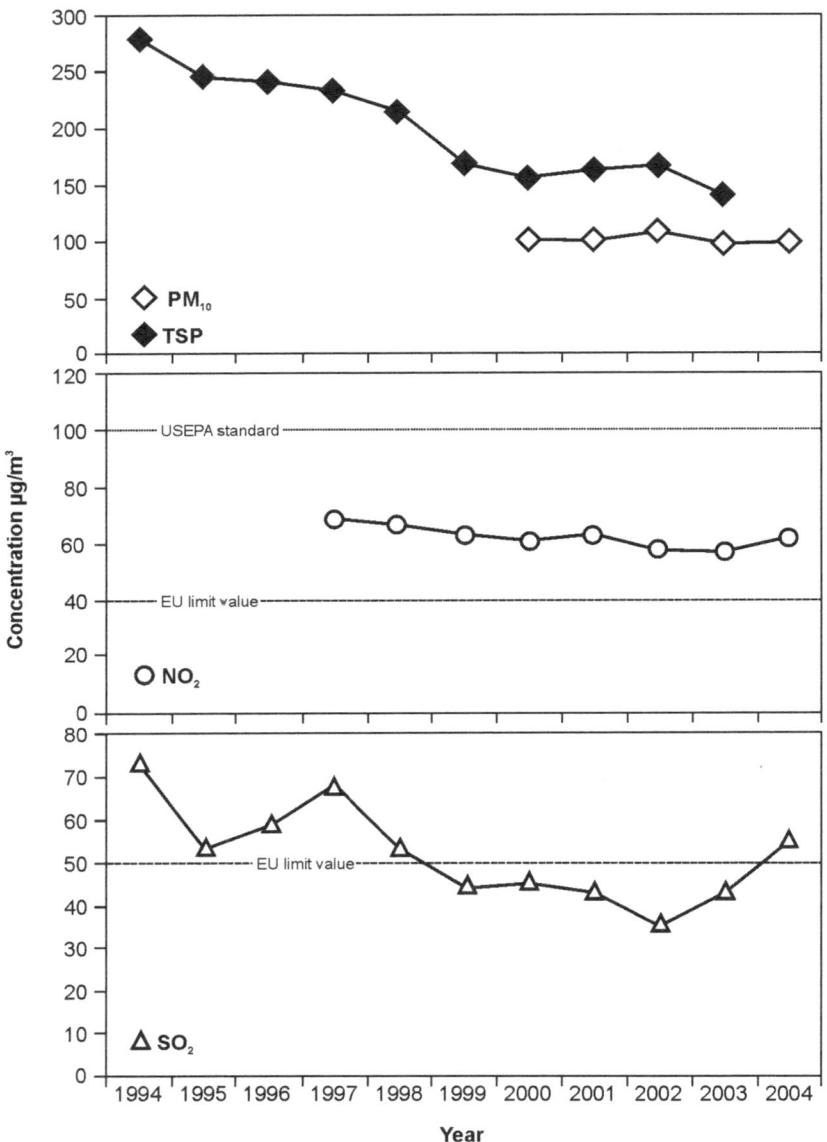

Figure 3.16 *Average annual concentrations of PM$_{10}$, TSP, NO$_2$, SO$_2$, Shanghai 1994–2004*

Source: Fu (2004)

2000 and 2004 annual levels of PM_{10} remained stagnant at approximately 100 $\mu g/m^3$, corresponding to the standard. Between 1997 and 2004 NO_2 levels slightly decreased. In contrast, SO_2 levels decreased substantially to practically half their 1994 values in 2002 but show an increasing tendency since then (Fu, 2004).

Health

With regard to the impact of air pollution on health, every 10 $\mu g/m^3$ increase over a 48-hour moving average of PM_{10}, SO_2 and NO_2, corresponds to a relative risk of non-accident mortality of 1.003, 1.016 and 1.020 (i.e. to 0.3, 1.6 and 2.0 per cent excess mortality), respectively, in Shanghai residents (Kan and Chen, 2003a; 2003b). In terms of effect on stroke mortality, an increase of 10 $\mu g/m^3$ of PM_{10}, SO_2 and NO_2 corresponds to a relative risk of 1.008, 1.017 and 1.029, respectively (Kan et al, 2004a). Each increase of 10 $\mu g/m^3$ in PM_{10}, SO_2 or NO_2 was found to correspond to a relative risk of diabetes mortality of 1.006, 1.011 or 1.013, respectively (Kan et al, 2004b).

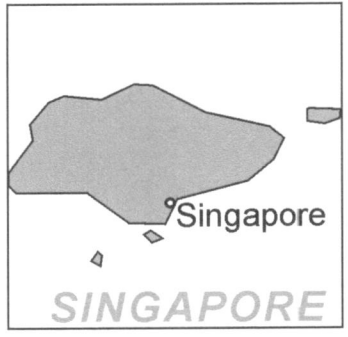

SINGAPORE

UN estimated population 2003: 4.25 million
UN projected population 2015: 4.71 million
Area: 697.1 km²
Climate: Tropical rain forest
Map reference: 1° 19'N, 103° 51'E
Annual mean precipitation: 2272.2 mm
Altitude: 10 m
Annual mean temperature range: 23–31°C

SITUATIONAL ANALYSIS

General

Singapore is located at the southernmost tip of the Malayan peninsula, between Malaysia and Indonesia. It comprises the island of Singapore and 63 islets within its territorial waters. The main island has a total area of 697.1 km². Singapore has a tropical rain forest climate and is 10 metres above sea level.

Legislation

The 1999 Environmental Pollution Control Act (EPCA) and its subsidiary regulations regulate air pollution. The 2000 Environmental Pollution Control (Air Impurities) Regulations stipulate emission standards for air pollutants which repeal the original 1978 Clean Air Standards. The EPCA consolidates previous laws on air, water and noise pollution and therefore provides a more comprehensive legislative framework for the control of environmental pollution (NEA, 2005). The USEPA air quality standards and the WHO air quality guidelines are used as guidelines in the evaluation of the ambient air quality.

Since January 2001 all petrol- and diesel-driven vehicles were required to comply with the Euro II emission standard. From 1 July 2003 all motorcycles/scooters were required to comply with the exhaust emission standard as specified in the European Directive 97/24/EC before they could be registered for use in Singapore. In addition to emission standards, all vehicles are also subject to periodic mandatory inspections.

The National Environment Agency (NEA) has gradually required the use of cleaner fuel, such as the removal of tetraethyl lead in petrol. The lead in petrol was phased out by 1 July 1998. To further reduce emissions from diesel vehicles, the permissible level of sulphur in diesel was also reduced from 0.3 per cent to 0.05 per cent by weight from 1 March 1999. This low-sulphur diesel has also helped reduce levels of SO_2 and particulate emissions from diesel-driven vehicles, allowing the introduction of more stringent Euro II emission standards for diesel-driven vehicles. The specifications for diesel fuel will be upgraded in preparation for the adoption of Euro IV standards for diesel vehicles in October 2006 (NEA, 2004a; 2005).

Emissions

Singapore does not have air emissions inventories since the authorities believe that their extensive air monitoring and enforcement programmes have been efficient in keeping the air quality of Singapore within international standards. The 2005 State of Environment reports that the main sources of air pollution in Singapore are from the burning of fossil fuel for heat generation in industries, electricity generation and transportation. The sources of air pollution then can be grouped into three categories as follows: stationary sources such as power stations, oil refineries and industries; mobile sources such as motor vehicles; and others such as transboundary air pollution.

Monitoring

The NEA monitors the ambient air quality through the telemetric air quality monitoring and management system. This comprises a network of remote air monitoring stations linked to a central control system via dial-up telephone lines. There are 16 remote air monitoring stations in the network, of which 14 stations monitor ambient air quality and two monitor roadside air quality (NEA, 2004a). The air monitoring stations are strategically located to accurately monitor the air quality at different parts of the island. Ambient air monitoring sites are classified as urban, industrial and suburban depending on the activities in the area where they are located. The two roadside stations are situated near busy roads or expressways and are used to assess the effectiveness of NEA's vehicle emission programme. Automatic analysers and equipment are installed at the stations to measure the concentrations of major air pollutants, such as SO_2, CO, O_3, PM_{10} and NO_x.

Since 2001 there has been a further decrease in SO_2 levels. In 2004, the annual average level was below 15 $\mu g/m^3$, considerably lower than the USEPA standard of 80 $\mu g/m^3$ and below the EU limit value of 20 $\mu g/m^3$ which is designed for protection of ecological systems. Annual average lead levels are below the WHO guideline value of 0.5 $\mu g/m^3$. Figure 3.17 presents the annual concentrations for PM_{10}, O_3, NO_2 and CO for the period 1994–2004.

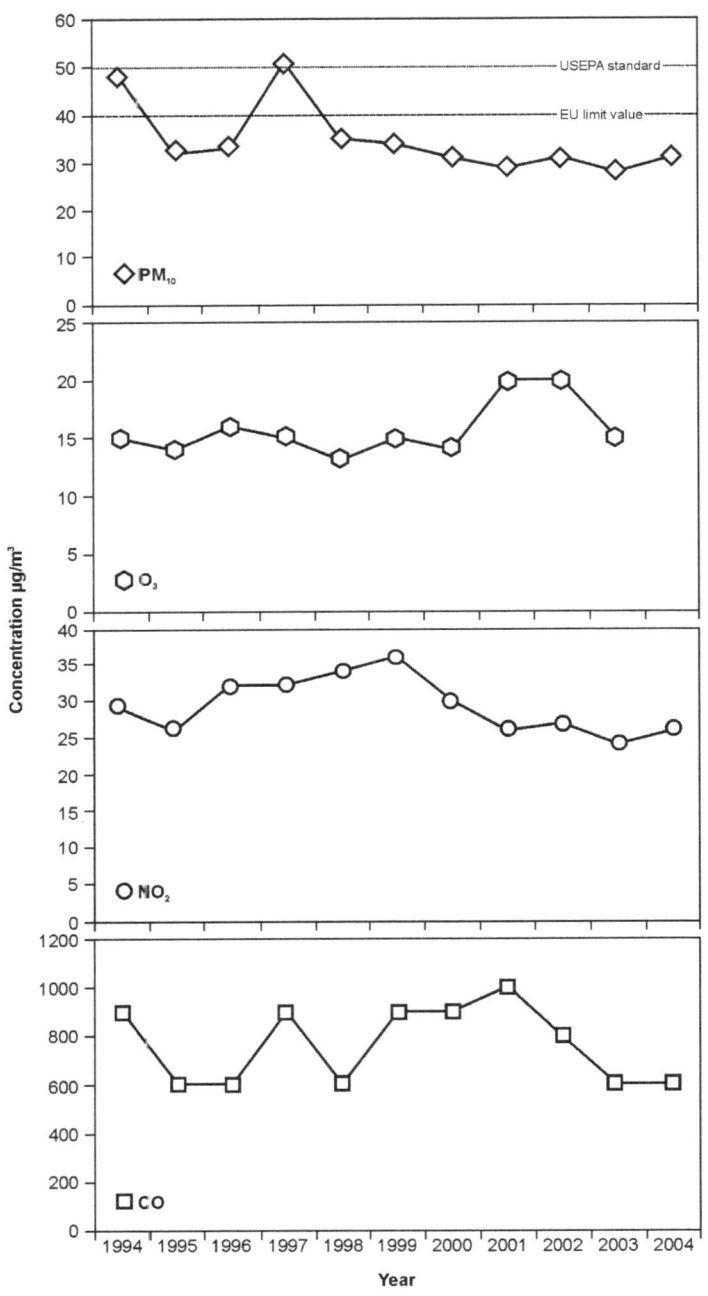

Figure 3.17 *Average annual concentrations of PM_{10}, O_3, NO_2, CO, Singapore 1994–2004*

Source: NEA (2004b)

PM is of specific concern in Singapore because of its impact on human health. Since 1998 PM_{10} annual levels have been below 35 $\mu g/m^3$ and comply with the USEPA standard of 50 $\mu g/m^3$ but exceed the WHO guideline value of 20 $\mu g/m^3$. Singapore is currently striving to meet the USEPA standard of 15 μgm^3 for $PM_{2.5}$, since black carbon and sulphates are big contributors to fine particulates in the city. Coarse PM in Singapore is caused mostly by sea salt.

Annual O_3 levels are low and 1-hour averages meet the USEPA standard, which stipulates that the one-hour O_3 concentration should not exceed 240 $\mu g/m^3$ on more than one occasion per year.

NO_2 concentrations varied between 24 and 36 $\mu g/m^3$ in the period 1994–2004 and comply with the USEPA standard (100 $\mu g/m^3$), and the EU limit and WHO guideline value (40 $\mu g/m^3$). In the same period, annual averages of 8-hour CO levels fluctuated between 600 and 1000 $\mu g/m^3$ complying with the USEPA standard and EU and WHO limits of 10,000 $\mu g/m^3$.

Health

A number of studies have addressed the impacts of air pollution in Singapore, particularly the effects of haze and PM on the health of residents. In 1997 Singapore was greatly affected by regional haze from the end of August to the first week of November. During this period there was a 30 per cent increase in outpatient attendance coinciding with haze episodes. The following increases were seen: 12 per cent upper respiratory tract illness, 19 per cent asthma and 26 per cent rhinitis were associated with the increase in PM_{10} levels from 50 $\mu g/m^3$ to 150 $\mu g/m^3$. There was also an increase in haze-related accidents. The 1997 haze episode is still considered as mild as there was no associated change in hospital admissions and mortality (Emmanuel, 2000).

Because of a high prevalence of asthma, Singapore is concerned about any factors that can affect asthma patients. One in five children has asthma (Chew et al, 1999a) and a further one in five reports suffering from asthma but has not been formally diagnosed (Goh et al, 1996). Linkages between ambient air pollution levels in the city and asthma exacerbation were established by a five-year time-series study. Despite ambient concentrations of SO_2, TSP, NO_2 and O_3 complying with USEPA standards and WHO guideline values, there was a positive correlation between levels of each of these pollutants and emergency department visits due to asthma for patients aged 3 to 12 years old. There was an increase of 2.9 ED visits for every 20 $\mu g/m^3$ increase in SO_2 concentration one day after days that levels were above 68 $\mu g/m^3$. Similarly, an increase of 5.8 ED visits for every 20 $\mu g/m^3$ increase in TSP concentration one day after days that levels were above 73 $\mu g/m^3$. The findings of the study suggested that asthmatic children are susceptible to increases in pollutant concentration despite the ambient levels still being within the accepted levels. The correlation, however, was not observed in children and young adults aged 13 to 21 years old (Chew et al, 1999b).

Although ambient air concentrations in Singapore have always met the international standards set by USEPA and guideline values derived by WHO before 2005, the economic cost of air pollution on human health in Singapore can still be considered significant. A study by the National University of Singapore has estimated that the total economic cost of particulate air pollution for Singapore is US$3662 million, which was approximately 4.31 per cent of GDP in 1999. This was based on a damage-function/dose response approach on morbidity and mortality effects (Quah and Boon, 2003).

Surabaya

INDONESIA

SURABAYA

UN estimated population 2003: 2.62 million
UN Projected population 2015: 3.45 million
Area: 326 km²
Climate: Tropical rain forest
Map reference: 7° 13'S, 112° 45'E
Annual mean precipitation: 1321 mm
Altitude: 3–10 m
Annual mean temperature range: 25.5–33°C

SITUATIONAL ANALYSIS

General

Surabaya is the second largest city in Indonesia after Jakarta and it is the capital of the east Java Province. It covers an area of 326 km² and is approximately 3–10 metres above sea level. Surabaya has a tropical rain forest climate.

Legislation

The 1997 Government Act No. 23 on Environmental Management and the 1999 Government Regulation No. 44 on Air Pollution Control are the main acts controlling air pollution in Indonesia. The 1999 Act sets out the mandate for setting up standards and acceptable practices in air pollution control for stationary and mobile sources (ADB, 2002). Surabaya's ambient air quality standards are based on the 1999 Government Decree No. 41 which established national ambient air quality standards but are less stringent than the standards adopted by Jakarta.

The SO_2 annual mean standard (60 µg/m³) is more stringent than the USEPA standard (80 µg/m³) and only slightly higher than the EU limit value (50 µg/m³). The standard for SO_2 for 24-hour exposure (365 µg/m³) is comparable to the USEPA standard but lenient compared to the WHO guideline (20 µg/m³) and EU limit value (125 µg/m³). The SO_2 1-hour standard of 900 µg/m³ is more than double the EU limit value of 350 µg/m³. The NO_2 1-hour (400 µg/m³) and annual standards (100 µg/m³) valid in Surabaya are also more lenient than WHO guideline and EU limit values (200 µg/m³ and 40 µg/m³, respectively). The annual NO_2 standard equals the USEPA standard. The O_3 standard for 1-hour exposure (235 µg/m³) is equivalent to the USEPA standard (240 µg/m³). The 24-hour PM_{10} standard (150 µg/m³) is three times the EU limit value (50 µg/m³) and more than seven times the WHO guideline value (20 µg/m³). The CO 1-hour exposure standard (30,000 µg/m³) is equivalent to both WHO and EU values and more stringent than the USEPA standard.

Surabaya follows Indonesia's Act No. 14 (1992) on Road Traffic and Transport, which states that to prevent air and noise pollution every motor vehicle must meet emission and noise standards. Emission tests are integrated into the roadworthiness test under the 1993 Government Regulation No. 43 on Vehicles and Motorists. There are two types of roadworthiness tests for vehicles in Indonesia: type approval tests for new type vehicles, and regular inspections for in-use vehicles that have passed the type approval test.

The 1993 Decree No. 35 sets motor vehicle exhaust emissions standards for Indonesia. The legislation stipulates the permissible limits for CO and HC for gasoline motorcycles, motor vehicles and black smoke for diesel vehicles. The CO and HC are measured at idle condition and smoke is measured at free acceleration.

Based on the 1995 Decree No. 13 five types of emission standards have been established for stationary sources of air pollution. The standards apply to the iron and steel, pulp and paper, cement and coal-fired power sectors, with all other industries grouped together as 'other industries'. All standards have been applied since May 1995 and stricter emission standards were introduced in 2000. Indonesia intends to improve the regulation to introduce stricter standards (GEF, 1998).

Emissions

There is no comprehensive emissions inventory or source apportionment of air pollution in Surabaya. One 2000 study shows emissions estimates for SO_2, HC, CO, NO_x and dust from refining industries, non-mobile sources and mobile sources based on fuel consumption (Silaban, 2003). Industry is a major source of dust or particulate emissions especially the refining industry, followed by the food industry, while mobile sources are major emitters of SO_2 and NO_x.

Monitoring

In 1999, Indonesia established a network of ambient air quality monitoring stations in ten cities network through the Indonesian Environmental Impact Management Agency (BAPEDAL). Five of these stations are located in Surabaya. The AQMS network monitors the concentrations of NO_2, HC, SO_2, PM_{10}, CO and O_3. In addition it records meteorological data which includes wind direction and speed, humidity, solar radiation and temperature (ADB, 2002).

Figure 3.18 presents the annual levels for PM_{10}, O_3, NO_2 and CO for the period 2001–2004. In the period 2001 to 2003 SO_2 concentrations complied with the annual

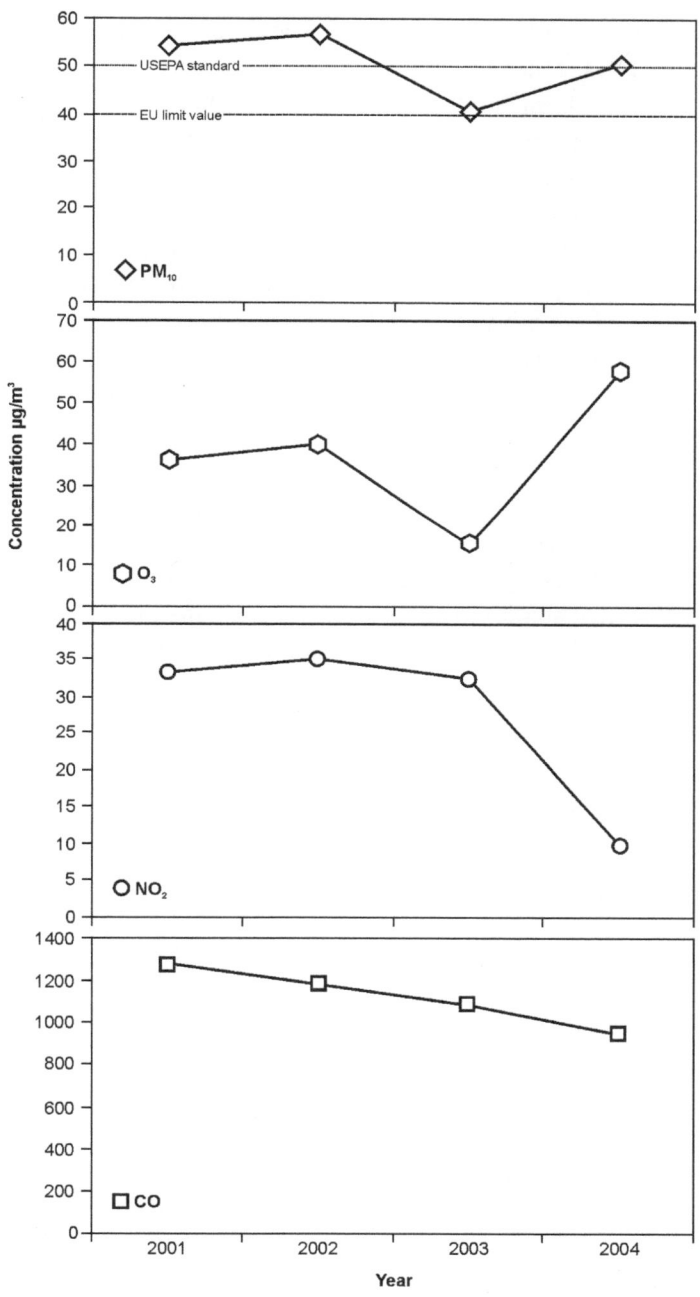

Figure 3.18 *Average annual concentrations of PM$_{10}$, O$_3$, NO$_2$, CO, Surabaya 2001–2004*

Source: Air Laboratory of Surabaya (2003)

standard (60 μg/m^3); however, in 2004 the standard was exceeded due to a rapid increase in annual SO$_2$ concentrations.

Annual PM$_{10}$ levels in Surabaya for 2001 to 2004 generally comply with the USEPA standard (50 μg/m^3). In the same period the observed O$_3$ annual levels also complied with the Indonesian standard for 1-year averaging (50 μg/m^3). Very low O$_3$ concentrations were observed in 2003 leading to the substantial decrease in that year which merits further clarification. Annual NO$_2$ concentrations in 2001–2004 are relatively low and comply with the standard (100 μg/m^3). During the same period 24-hour levels of CO comply with Indonesia's 24-hour averaging (10,000 μg/m^3) standards and show a decreasing tendency.

Health

No reports on the health impacts of air pollution in Surabaya appear to exist in the international literature. There are no official cost figures with regard to the environmental and health effects of air pollution in Surabaya due to air pollution. There are no available surveys or studies available which determine the cost and the severity of the magnitude and significance of the environmental effects.

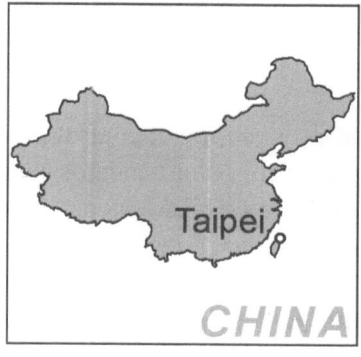

TAIPEI

UN estimated population 2002: 2.51 million
UN projected population 2010: 2.45 million
Area: 271.8 km²
Climate: Humid sub-tropical
Map reference: 25° 04'N, 121° 33'E
Annual mean precipitation: 2161.9 mm
Altitude: 8 m
Temperature range: 15–26.7° C

SITUATIONAL ANALYSIS

General

Taipei city lies on the island of Taiwan in a basin of land covering an area of 271.8 km². It is surrounded by mountains and is 8 metres above sea level. In 2001 the designated planning area of the city was limited to approximately 49.29 per cent of its total area. This is mainly due to its mountainous topography (DUD, 2003). Taipei has a humid sub-tropical climate.

Legislation

The 1975 Air Pollution Control Act (APCA), which was revised in 2002, provides the legal basis for air pollution-related laws and regulations. The Act defines the different responsibilities and tasks of central and local government agencies. The Environmental Protection Agency, in consultation with the local competent authorities, is responsible for air pollution management, such as programme initiation, establishing emission standards and guidelines on fees. Local government agencies may draft more stringent emission controls but these require approval by the central authority (EPA, 2002). The 2003 Air Pollution Control Act Enforcement Rules describe in detail the specific responsibilities of the competent authorities as well as the specific implementation of air quality monitoring (e.g. types of stations, pollutants to be measured and reporting).

The Department of Environmental Protection (DEP) outlines the air quality standards for Taipei. Annual average levels of ambient concentration of PM_{10} and NO_2 must not exceed the maximum limit of 65 µg/m³ and 95 µg/m³, respectively, while 8-hour average limits for CO and O_3 are set at 10,000 µg/m³ and 120 µg/m³ respectively. Monthly average levels for lead should not exceed 1.0 µg/m³ (DEP,

2004). All other Taiwan standards are equal to those of the USEPA. The PM_{10} and NO_2 annual Taiwan standards are lenient compared with the EU limit values (both 40 $\mu g/m^3$, respectively).

Taiwan has established a relatively extensive motorcycle emission control programme both for new and in-use motorcycles. Type approval emission standards have been set for CO, HC and $HC + NO_x$. If motorcycles fail to meet the EPA standards, the burden is not limited only to the owners but also the manufacturers. If the motorcycles have been properly maintained yet fail to meet the standards, manufacturers are obliged to recall and repair them. A fine of approximately US$47 is imposed if motorcycles fail the roadside inspections and approximately US$94 for failure of periodic idle tests (Chuang, 2001).

Since 2000 gasoline in Taipei has been lead-free (ADB, 2003). Currently, gasoline in the market has to contain only up to 1 per cent benzene (by volume) and 180 ppm of sulphur, while diesel has to meet the limit of 0.035 per cent (350 ppm) by weight of sulphur content. In 2007 more stringent fuel standards will be implemented in Taiwan to make fuel compatible with Taiwan's phase 4 emissions standards for diesel automobiles which are in line with the international trend towards low sulphur levels. Sulphur content in diesel and gasoline will be limited to a maximum of 50 ppm in 2007 (EPA, 2003a).

The 1986 Toxic Chemical Substances Control Act (revised in 1999) prevents any toxic chemical substances from polluting the environment or from endangering human health. The Act outlines the responsibilities of national and local governmental offices. It classifies toxic substances according to their level of impact: Class I includes toxic chemical substances that are not prone to decompose in the environment or that pollute the environment or endanger human health due to bioaccumulation, bioconcentration or biotransformation. Class II includes toxic chemical substances that cause tumours, infertility, teratogenesis, genetic mutations or other chronic diseases. Class III includes toxic chemical substances that endanger human health or the lives of biological organisms immediately upon exposure. Class IV includes toxic chemical substances for which there is concern of pollution of the environment or a risk to human health (EPA, 2003b).

The APCA requires industries or establishments to secure the necessary permits from the responsible agency at the municipal or county (city) government level before installation, operation or modification of any stationary source. Before a permit is granted, those applying for the permit need to first undertake an environmental impact assessment and submit an Air Pollution Control Plan. Even after a permit is secured, the concerned establishments or stationary sources are still subject to reviews and fines from the responsible agency in the course of actual operation (EPA, 2002; 2003c).

Emissions

A 2003 inventory of pollutant emission sources in the city shows that combustion, road transport and fugitive emissions contribute greatly to the total emission load of most pollutants. Fugitive emissions contribute as much as 85 per cent of total TSP emissions and 50 per cent of PM_{10}. Emissions from petroleum-fuelled vehicles are also a key source of NO_x (26 per cent), non-methane hydrocarbons (13 per cent) and CO (51 per cent), while diesel vehicles are a key source of SO_x (16 per cent) and NO_x (38 per cent). Motorcycles in the city are major contributors of NMHCs (21 per cent) and CO (38 per cent) (BEP, 2003).

Monitoring

The Taiwan Air Quality Monitoring Network (TAQMN) was established in 1990 and initially had 19 air quality monitoring stations. These were increased to 66 stations in 1993 and 72 stations in 1998. Pollutants being monitored under the TAQMN include NO_2, CO, PM, O_3, SO_2 and HCs (EPA, 2003d). Taipei is located within the northern air quality control basin where the EPA has installed 19 continuous emissions monitoring stations. Eight of these automatic and computerized air quality monitoring stations are located in Taipei city, five of which are dedicated for ambient monitoring (DEP, 2003).

Figure 3.19 presents the average annual concentrations for PM_{10}, O_3, NO_2 and CO. With the exception of O_3 there is a decreasing tendency in the annual concentrations of all pollution monitored. Annual SO_2 concentrations are below 20 µg/m^3 in Taipei and comply with the Taiwan standard. For the period 1994 to 2004 annual PM_{10} concentrations varied between 40 and 60 µg/m^3 and complied with the Taiwan standard of 65 µg/m^3. There has been a slightly increasing trend in O_3 concentrations between 1994 and 2003. O_3 annual levels fluctuated between 140 and 170 µg/m^3 in 1994–2003 with a sharp decrease below 50 µg/m^3 in 2004. Annual NO_2 and CO levels have also decreased with a substantial decrease in CO in 2004 to 1043 µg/m^3. With the exception of O_3, annual average concentrations of all pollutants monitored in Taipei during 1994–2004 met their respective air quality standards (DEP, 2003; 2004).

Health

Several studies on health impacts exist in Taipei. One study indicated that the higher levels of ambient pollutants (PM_{10}, NO_2, CO and O_3) increase the risk of hospital admissions for cardiovascular diseases especially during warm days – temperatures greater than or equal to 20°C (Chang et al, 2005).

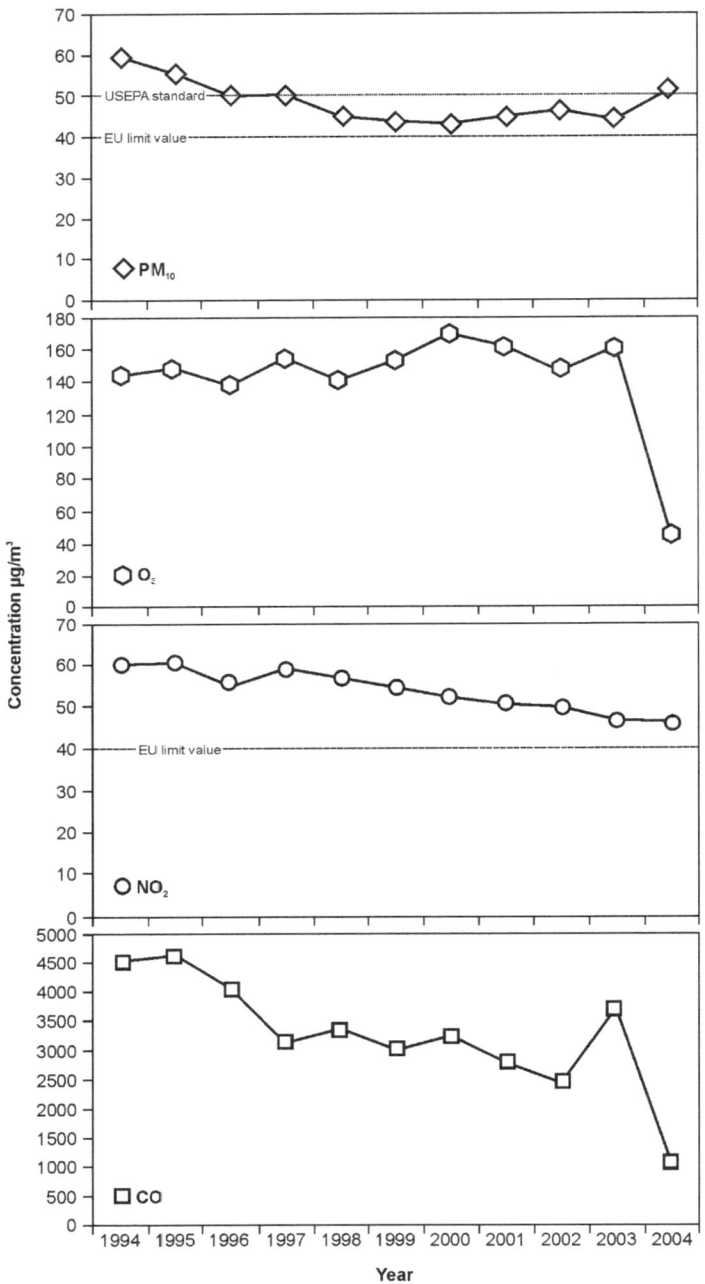

Figure 3.19 *Average annual concentrations of PM$_{10}$, O$_3$, NO$_2$, CO, Taipei 1994–2004*

Source: DEP (2004)

Dust storms originating from mainland China caused an approximately 8 per cent increase in risk of respiratory disease one day after the storm, 5 per cent increase in risk of total deaths two days after the event as well as a 2.6 per cent increased risk for circulatory disease two days following the storm (Chen et al, 2004). Another study observed a significant relationship between air pollution and daily mortality due to respiratory diseases in Taipei (Yang et al, 2004).

A study of the short-term effect of O_3 on the pulmonary function of schoolchildren revealed that the decrease in children's lung function was statistically significant if the peak hourly O_3 concentrations are higher than 80 parts per billion (Chen et al, 1999). In an earlier study, Chen et al (1998) showed that children living in urban areas with elevated levels of air pollution had consistently higher rates of symptoms and diseases than those from rural areas.

Individual levels of SO_2, NO_x and NO_2 were positively correlated with the frequency of absence of primary school students due to illness while PM_{10} and O_3 did not show any correlation with absence frequency (Hwang et al, 2000). A study of the association between prevalence of asthma in middle school students, air pollution and weather confirms that asthma prevalence rates were highly correlated with certain traffic-related air pollutants (CO and NO_x) but not with other pollutants such as PM_{10}, O_3 and SO_2 (Guo et al, 1999).

TOKYO

UN estimated population 2003: 35 million
UN projected population 2015: 36.2 million
Area: 2187 km²
Climate: Humid sub-tropical
Map reference: 35° 41'N, 139° 48'E
Annual mean precipitation: 1460 mm
Altitude: 220 m
Annual average temperature range: –2–30° C

SITUATIONAL ANALYSIS

General

Tokyo is the capital city of Japan and is situated in the centre of the Japanese archipelago. Tokyo covers an area of 2187 km², which accounts for approximately 0.6 per cent of Japan's total land area. It is approximately 220 metres above sea level and has a humid sub-tropical climate.

Legislation

The 1968 Air Pollution Control Law (amended in 1996) is the key legislation related to air pollution in Japan. Under this law all sectors have a responsibility for ensuring the improvement of air quality in the country. With the cooperation of local governments, the state is required to ensure that all sectors are informed of the state of air quality in the country and that the analysis of information and relevant research on air quality is publicly available. Corporations are required to ensure that they are fully aware of any emissions of hazardous air pollutants and are able to take the necessary measures to reduce these discharges. A specific section of the law is devoted to making citizens responsible for controlling emissions of pollutants associated with their daily activities (MOE, 2005a).

The Tokyo Metropolitan Government implements national ambient air quality and motor vehicle exhaust emission standards under the Basic Environment Law and the Air Pollution Control Law. Japanese standards are to some extent more stringent than USEPA standards. An exception is the CO standard (22,000 µg/m³) for 8-hour exposures which is more than double the respective USEPA standard, EU limit value and WHO guideline value (10,000 µg/m³). The 24-hour standard for SO_2 (106 µg/m³) is more stringent than the USEPA standard (365 µg/m³), the EU limit (125 µg/m³) and is lenient compared to the WHO guideline value (20 µg/m³). The 24-hour NO_2

standard range (75–113 µg/m³) has no WHO, EU or US counterpart. The O_3 1-hour standard of 120 µg/m³ is half the USEPA standard value. The Japanese PM_{10} standard for 24-hour exposures (100 µg/m³) is lenient compared to the EU limit and WHO guideline value of 50 µg/m³ but more stringent than the USEPA standard (150 µg/m³). Annual standards for SO_2, NO_2 and PM_{10} do not appear to have been adopted in Japan (MOE, 2005b).

Regulation of air pollution from mobile sources is a joint responsibility of the MOE and the Ministry of Land, Infrastructure and Transport. Since the 1980s, when vehicle emission standards were first introduced in Japan, the standards have undergone a number of amendments. Further changes were made in 2003 and scheduled for implementation in 2005. Once implemented, it is thought that these standards will have the most stringent diesel emission limits in the world (Dieselnet, 2005a).

Aside from the vehicle emission limits stipulated by the air pollution law, there is the 1992 law concerning special measures to reduce the total amount of NO_x emitted from motor vehicles in specified areas. The law is aimed at eliminating old, polluting vehicles from in-use fleets in selected locations which includes Tokyo (MOE, 2005b). The regulation was amended in June 2001 by the automotive and PM Law to tighten the existing NO_x requirements and to add PM control provisions. With this law, vehicles are given a specific life span depending on the type and weight and are also required to be retrofitted with NO_x and PM control devices (Dieselnet, 2005b).

In December 2000 the Ordinance on Environmental Preservation was drafted in Tokyo. This ordinance includes a diesel emission control regulation (retrofit programme); a vehicle environmental management plan for businesses with more than 30 vehicles; the promotion of low emission vehicles and an idling-stop practice (TMG, 2004).

Japanese fuel standards have experienced a number of improvements over time starting with 2000 ppm sulphur content in diesel to a mandatory 50 ppm in 2005 and 10 ppm in 2007. However, low-sulphur diesel fuels are being voluntarily introduced by Japanese petroleum suppliers earlier than mandated. Diesel fuel with 10 ppm sulphur content has been readily available in the market since January 2005 (Dieselnet, 2005b).

Emissions from stationary and area sources in Tokyo are no longer major contributors to air pollution compared to mobile sources. The strict implementation of the national regulatory measures against air pollutants emitted from factories and business sites has resulted in the low contribution to pollution levels from this sector. Regulation of stationary and area sources include those for sulphur oxides, soot and dust, harmful substances such as cadmium (Cd), cadmium compounds, chlorine (Cl_2), hydrogen chloride (HCl) fluorine (F), hydrogen fluoride (HF), lead, lead compounds, NO_x and more specific substances such as benzene, trichloroethylene and tetrachloroethylene, among others (MOE, 2005b).

Emissions

In Tokyo mobile sources are a main contributor to emissions of NO_x and PM_{10}. In 2000, mobile sources accounted for approximately 52 per cent of NO_x emissions and 64 per cent of PM_{10} emissions (BoE, 2005).

Monitoring

Tokyo has 24-hour monitoring devices at 82 strategic locations, 47 of which are for continuous ambient air quality monitoring and 35 specifically for evaluating roadside pollution. Pollutants monitored include NO_2, NO, PM_{10}, photochemical oxidants, SO_2 and CO. Wind direction and velocity, temperature and humidity and other pollutants such as methane and non-methane HCs are also measured simultaneously (BoE, 2005).

Figure 3.20 presents the annual levels for PM_{10}, O_3, NO_2 and CO in Tokyo. In the period 1996–2004 there has been a decreasing tendency in the concentrations of SO_2, NO_2, PM_{10} and CO. O_3 annual levels appear to be increasing. Annual PM_{10} levels are approaching 30 $\mu g/m^3$ while CO levels are low. However, the decrease in NO_2 levels has been relatively small. Therefore NO_2 and O_3 remain the main pollutants of concern. Due to the lack of long-term standards in Japan, a comparison of annual levels with standards is not possible.

Health

In 2001 the US National Institute of Environmental Health Sciences and the National Institute for Environmental Studies in Japan undertook a health study to examine the effects of temperature and air pollutants on cardiovascular and respiratory diseases in males and females aged 65 and above. This study demonstrated that a strong association exists between concentrations of NO_2 and PM_{10} and emergency cases of the following diseases: myocardial infarction, asthma, acute and chronic bronchitis, pneumonia, angina and cardiac insufficiency. Data on hospital emergency incidences (expressed as incidence per million persons) of cardiovascular and respiratory diseases were related to maximum temperatures and pollution levels of NO_2, O_3, PM_{10}, SO_2 and CO. The study covered the months of July and August (when temperatures are usually at their highest) for the period 1980–1995 (Ye et al, 2001).

Two years earlier, in 1999, the same organizations also conducted a similar study on risk factors for heat stroke. The same months of July and August for the period 1980–1995 were examined. However, this time three age groups (0–14, 15–64 and > 65) were considered. The months chosen are especially significant as 85 per cent of heat stroke cases occur in July or August. Heat stroke incidences reported at four

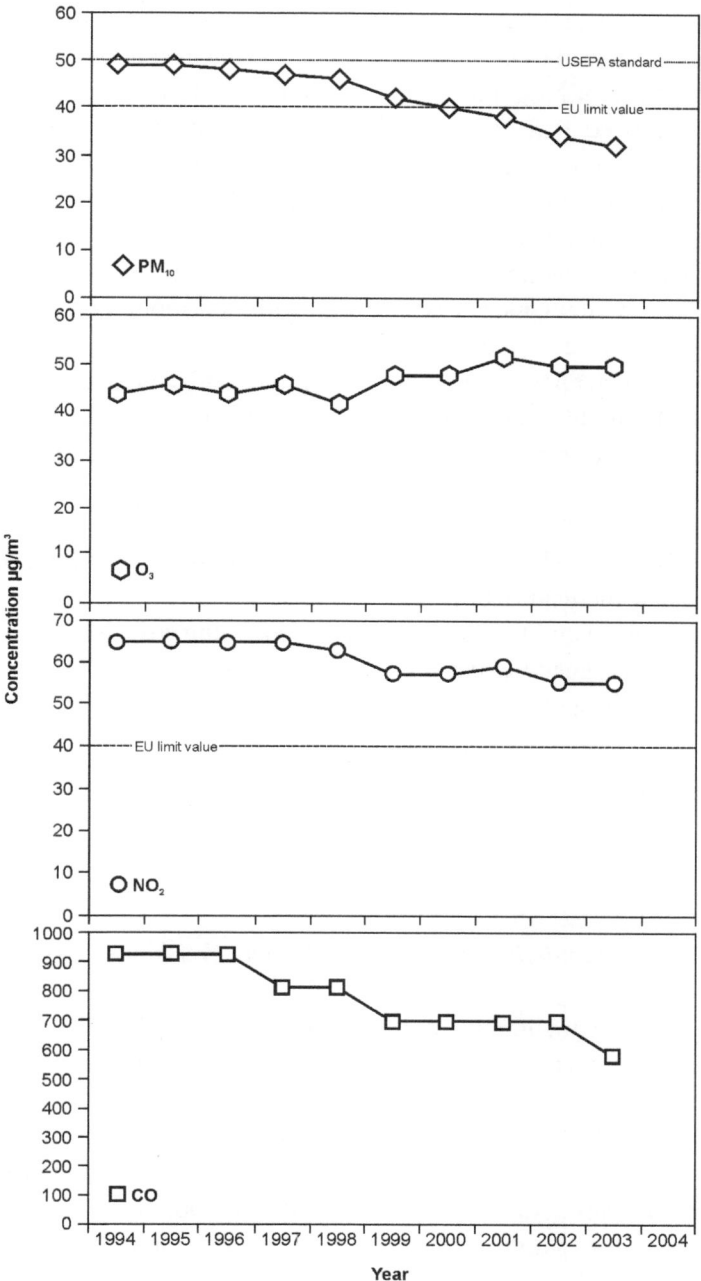

Figure 3.20 *Average annual concentrations of PM$_{10}$, O$_3$, NO$_2$, CO, Tokyo 1994–2004*

Source: BoE (2005)

hospitals in Tokyo were classified according to gender and age. Males of all ages suffered more heat strokes compared to females. Linear regression analyses of heat stroke as a function of NO_2, PM_{10} and the combination of PM_{10}, NO_2 and maximum temperatures showed that NO_2 and maximum temperatures were significant risk factors for all age groups and gender while PM_{10} was not (Piver et al, 1999).

Sekine et al (2004) investigated the long-term effects of exposure to motor vehicle pollution on the pulmonary function. A total of 733 female adults were subjected to a test of pulmonary function parameters: forced expiratory volume in one second (FEV_1) and forced vital capacity (FVC). Air pollution monitored included NO_2 and PM. The subjects were divided into three groups according to their exposure to air pollution. Prevalence rates of respiratory symptoms and decreases in FEV_1 were most significant in the highly exposed group. Long-term exposure to NO_2 and PM was related to an increase in airway resistance and a chronic decrease in pulmonary lung function.

Voorhees et al (2000) estimated the benefits and costs of past NO_2 control policies as medical expenses and lost work time, using environmental, economic, political, demographic and medical data from 1973 to 1994. Direct costs were calculated as annualized capital expenditures and one year's operating costs for regulated industries plus governmental agency expenses. The best net estimate of the avoided medical costs and avoided wage losses amounted to approximately US$14 billion.

REFERENCES

All cities

ADB (2003) *Policy Guidelines for Reducing Vehicle Emissions in Asia*, Manila, Asian Development Bank

APCEL (2006) 'Vietnam Implementing Regulations', Asia-Pacific Centre for Environmental Law, http://sunsite.nus.edu.sg/apcel/dbase/vietnam/regs.html accessed in April 2006

AQMP (2003a) *Air Quality Management Project*, Inter-Ministerial Committee Meeting. Working Paper 2 – Revised Ambient Air Quality Standards for Bangladesh, Bangladesh, Department of Environment, August 2003

AQMP (2003b) *Air Quality Management Project*, Inter-Ministerial Committee Meeting. Working Paper 1 – Revised Vehicle Emission Standards for Bangladesh, Bangladesh, Department of Environment, August 2003

Ghose M. K. (2002) 'Controlling of Motor Vehicle Emissions for a Sustainable City', *TERI Information Digest on Energy and Environment (TIDEE)*, vol 1, no 2, June 2002, www.terienvis.nic.in/tidee1-2.pdf accessed in April 2006

Le, Van Khoa (2003) 'Air Quality Management in Ho Chi Minh City', www.iges.or. jp/kitakyushu/Meetings/Thematic%20Seminar/UAQM/Presentations/Ho%20 Chi%20Minh/Ho%20Chi%20Minh.pdf accessed in March 2006

Lee, J.-T. and Schwartz, J. (1999) 'Reanalysis of the Effects of Air Pollution and Daily Mortality in Seoul, Korea: A Case-crossover Design', *Environmental Health Perspectives*, vol 107, pp633–636

Malé Declaration (2000a) 'Baseline Information and Action Plan', United Nations Environment Programme Regional Resource Centre for Asia and Pacific, www. rrcap.unep.org/issues/air/maledec/baseline/indexsri.html accessed in February 2006

Malé Declaration (2000b) 'Chapter 5. National Response to Air Pollution Problem', www.rrcap.unep.org/issues/air/maledec/baseline/Baseline/India/INCH5.htm accessed in February 2006

MOE (2006) '2005 Environmental Statistics Yearbook', Ministry of Environment, Seoul, Republic of Korea http://eng.me.go.kr/docs/common/common_list.html? mcode=20&classno=10&topmenu=EMoEF (2002) 'Ministry of Environment and Forests', http://envfor.nic.in:80/legis/air/air1.html accessed in April 2006

MoEF (1981a) 'The Air (Prevention and Control of Pollution) Act, 1981', Ministry of Environment and Forests, New Delhi, India, http://envfor.nic.in:80/legis/air/air1. html accessed in February 2006

MoEF (1981b) 'The Air (Prevention and Control of Pollution) Act, 1981', New Delhi, India, Ministry of Environment and Forests, http://envfor.nic.in:80/legis/air/air1. html accessed in February 2006

MoEF (2004a) 'The Air (Prevention and Control of Pollution) Rules, 1983', Ministry of Environment and Forests, http://envfor.nic.in:80/legis/air/air3.html, accessed in February 2006

MoEF (2004b) 'Legislation on Environment, Forests and Wildlife', Ministry of Environment and Forests, http://envfor.nic.in/legis/legis.html, accessed in February 2006

Nguyen, Van Thai (2004) 'Strengthening Institutional Coordination for Air Quality Improvement in Vietnam', Presentation at the Cleaner Vehicles and Fuels in Vietnam Workshop, 13–14 May, 2004, Hanoi, Vietnam, Vietnamese Ministry of Transportation and US-AEP, www.nea.gov.vn/SukienNoibat/Chatluong_KK/hoinghi-hoi thao/Presentations%20ENGLISH/Mr.%20Tai%20Air%20quality%20coordinatio n %20Day%202.pdf accessed April 2006

WHO (2000) *Guidelines for Air Quality*, WHO/SDE/OEH/00.02, Geneva, World Health Organization, http://whqlibdoc.who.int/hq/2000/WHO_SDE_OEH_00.02_ pp1-104.pdf, http://whqlibdoc.who.int/hq/2000/WHO_SDE_OEH_00.02_pp105-190.pdf accessed in February 2006

WHO (2005) 'WHO Air Quality Guidelines Global Update 2005', Report of a working group meeting, Bonn, Germany, 18–20 October 2005, Geneva, World Health Organization, www.cepis.ops-oms.org/bvsea/fulltext/guidelines05.pdf accessed in April 2006

World Bank (2004a) 'For a Breath of Fresh Air: Ten Years of Progress and Challenges in Urban Air Quality Management in India (1993–2002)', Environment and Social Development Unit, South Asia Region, World Bank. India, http://lnweb18.world bank.org/SAR/sa.nsf/General/2F391E72031478F685256B17006FF5BB?Open Document accessed in February 2006

World Bank (2002b) *Philippine Environment Monitor: Let's All Act to Clean the Air*, Country office, Manila, The World Bank Group

Introduction

EU (1999) 'Council Directive 1999/30/EC of 22 April 1999 relating to Limit Values for Sulphur Dioxide, Nitrogen Dioxide and Oxides of Nitrogen, Particulate Matter and Lead in ambient air', *Official Journal of the European Union*, L 163, 29.06. 1999, pp41–60. Brussels, http://europa.eu.int/eur-lex/lex/LexUriServ/LexUriServ. do?uri=CELEX:31999L0030:EN:HTML accessed in February 2006

EU (2000) 'Directive 2000/69/EC of the European Parliament and of the Council of 16 November 2000 relating to Limit Values for Benzene and Carbon Monoxide in Ambient Air', *Official Journal of the European Union*, L 313, 13.12.2000, pp12–36. Brussels, http://europa.eu.int/eur-lex/lex/LexUriServ/LexUriServ.do?uri =CELEX:32000L0069:EN:HTML accessed in February 2006

EU (2002) 'Directive 2002/3/EC of the European Parliament and of the Council of 12 February 2002 relating to Ozone in Ambient Air', *Official Journal of the European Union*, L 67, 09.03.2002, pp14–30, Brussels http://europa.eu.int/eur-lex/pri/en/oj/ dat/2002/l_067/l_06720020309en00140030.pdf accessed in February 2005

The Times (2003) 'Climate Types and Extremes', *The Comprehensive Atlas of The World,* 11th edn, London, The Times, pp36–37, Index, 6–223

UN (2004) 'World Urbanization Prospects: The 2003 Revision', ST/ESA/SER. A/237, United Nations, Department of Economic and Social Affairs, Population Division, New York, www.un.org/esa/population/publications/wup2003 accessed in February 2006

USEPA (1997) 'National Ambient Air Quality standards (NAAQS)', Washington DC, US Environmental Protection Agency, Office of Air Quality Planning and Standards, Air and Radiation, www.epa.gov/air/criteria.html accessed in February 2006

Bangkok

Jinsart W., Tamura K., Loetkamonwit S., Thepanondh, S., Karita, K. and Yano, E. (2002) 'Roadside Particulate Air Pollution in Bangkok', *Journal of Air Waste Management Association*, vol 52, pp1102–1110

Mottershead, T. (2002) 'Thailand', in Mottershead, T. (ed.) *Environmental Law and Enforcement in the Asia-Pacific Rim*, Hong Kong, Singapore, Malaysia, Sweet and Maxwell

Oanh, N. T. K. and Zhang, B. (2004) 'Photochemical Smog Modeling for Assessment of Potential Impacts of Different Management Strategies on Air Quality of the Bangkok Metropolitan Region, Thailand', *Journal of Air Waste Management Association*, vol 54, pp1321–1338

PCD (2000) 'Air Emission Source Database Update 2000', Bangkok, Thailand, Pollution Control Department, www.pcd.go.th/ accessed in January 2005

PCD (2005) Personal Communication, Janejob Suksob, Planning and Processing Branch, Air Quality and Noise Management Bureau, Pollution Control Department, Bangkok, Thailand, 22 November 2005

Radian International LLC (1998) *Particulate Matter Abatement Strategy for the Bangkok Metropolitan Area*, Bangkok, Thailand, Pollution Control Department, Ministry of Science, Technology and Environment

Tamura, K., Jinsart, W., Yano, E., Karita, K. and Boudoung, D. (2003) 'Particulate Air Pollution and Chronic Respiratory Symptoms among Traffic Policemen in Bangkok', *Archives of Environmental Health*, vol 58, pp201–207

UNEP/BMA (2004) 'Bangkok State of the Environment 2003', United Nations Environment Programme – Asia Pacific Region, Bangkok Metropolitan Administration, www.rrcap.unep.org/reports/soe/bangkoksoe.cfm accessed in February 2006

Wongsurakiat, P., Maranetra, K. N., Nana, A., Naruman, C., Aksornint, M. and Chalermsanyakorn, T. (1999) 'Respiratory Symptoms and Pulmonary Function of Traffic Policemen in Thonburi', *Journal of the Medical Association of Thailand*, vol 82, pp435–443

World Bank (2002) 'Thailand Environment Monitor 2002 for Air Quality', Washington DC, World Bank, www.worldbank.org/ accessed in February 2006

Beijing

BJEPB (2005) *Annual Environmental Bulletin of Beijing*, Beijing Environmental Protection Bureau, www.bjepb.gov.cn (in Chinese) accessed in February 2006

Chang, G., Pan, X., Xie, X. and Gao, Y. (2003) 'Time-series Analysis on the Relationship between Air Pollution and Daily Mortality in Beijing' (in Chinese), *Wei Sheng Yan Jiu*, vol 32, pp565–568

Dan, M., Zhuang, G., Li, X., Tao, H. and Zhuang, Y. (2004) 'The Characteristics of Carbonaceous Species and their Sources in PM$_{2.5}$ in Beijing', *Atmospheric Environment*, vol 38, pp3443–3452

Kan, H. D., Chen, B. H., Fu, C. W., Yu, S. Z. and Mu, L. N. (2005) 'Relationship between Ambient Air Pollution and Daily Mortality of SARS in Beijing', *Biomedical and Environmental Sciences*, vol 18, pp1–4

SEPA (2006) 'Environmental Standards', Beijing, China, State Environmental Protection Agency, www.zhb.gov.cn/english/chanel-5/chanel-5-end-2.php3?chanel=5&column=1 accessed February 2006

Sun, Y., Zhuang, G., Wang, Y., Han, L., Guo, J., Dan, M., Zhang, W., Wang, Z. and Hao, Z. (2004) 'The Air-borne Particulate Pollution in Beijing – Concentration, Composition, Distribution and Sources', *Atmospheric Environment*, vol 38, pp5991–6004

Wang, X., Ding, H., Ryan, L. and Xu, X. (1997) 'Association between Air Pollution and Low Birth Weight: A Community Based Study', *Environmental Health Perspectives*, vol 105, no 5, pp514–520

Xu, X., Ding, H. and Wang, X. (1995a) 'Acute Effects of TSP and SO$_2$ on Preterm Delivery: A Community Based Cohort Study', *Archives of Environmental Health*, vol 50, no 6, pp407–415

Xu, X., Christiani, D. C., Li, B. and Huang, H. (1995b) 'Association of Air Pollution with Hospital Outpatient Visits in Beijing', *Archives of Environmental Health*, vol 50, no 3, pp214–220

Xu, X., Gao, J., Dockery, D. W. and Chen, Y. (1994) 'Air Pollution and Daily Mortality in Residential Areas of Beijing, China', *Archives of Environmental Health*, vol 49, no 4, pp216–222

Yu, T. (2002a) 'Air Quality Monitoring in Beijing', Presentation at Better Air Quality 2002 Workshop, Hong Kong, 16–18 December 2002, www.cse.polyu.edu.hk/~activi/BAQ2002/BAQ2002_files/Proceedings/Subworkshop2/sw2a-4TongYu.pdf accessed in February 2006

Yu, T. (2002b) 'Beijing Municipal Environmental Monitoring Center', presentation at the Energy and Environment Workshop (Energy Use, Spatial Planning and Air Quality), 12–14 November 2002, Shanghai

Zhang, Y., Zhu, X., Slanina, S., Shao, M., Zeng, L., Hu, M., Bergin, M. and Salmon, L. (2004) 'Aerosol Pollution in Some Chinese Cities', *Pure Applied Chemistry*, vol 76, no 6, pp1227–1239

Busan

APMA/KEI (2002) *International Joint Research Project on Air Pollution in the Megacities of Asia*, Seoul, Korea, Ministry of Environment

Busan (2001) *2000 Investigation of Vehicle Traffic Volumes (in Korean)*, Busan, Korea, The Busan Metropolitan Government

Busan (2002) *Implementation Plan for Air Quality Improvement in Busan City*, Busan, Korea, The Busan Metropolitan Government

Lee, J.-T., Shin, D. and Hung, Y. (1999) 'Air Pollution and Daily Mortality in Seoul and Ulsan, Korea', *Environmental Health Perspectives*, vol 107, pp149–154

Lee, J.-T., Kim, H., Hong, Y.-C., Kwon, H.-J., Schwartz, J. and Christiani, D. C. (2000) 'Air Pollution and Daily Mortality in Seven Major Cities of Korea, 1991–1997', *Environmental Research*, vol A 84, pp247–254

MOE (2002) 'Social Costs of Air Pollution were Estimated about 45 Trillion Won', *Environmental News*, 8 April, 2002, Seoul, Korea, Ministry of Environment

Colombo

AirMAC (2004) Personal Communication, Sugath Yalegama, General Secretary, Air Resources Management Centre, 16 July 2004

Chandrasiri, S. (1999) 'Controlling Automotive Air Pollution: The Case of Colombo City', Economy and Environment Program for Southeast Asia (EEPSEA) Research Report, http://web.idrc.ca/en/ev-8421-201-1-DO_TOPIC.html accessed in February 2006

Clean Air Sri Lanka Net (2006) Personal Communication, Ruwan Weerasooriya, Moderator, Clean Air Sri Lanka Net, 18 January 2006

Island, The (2003) 'The Deadly Danger of Urban Air Pollution', *The Island*, Sunday, 18 May 2003, www.southasianmedia.net/index_opinion3.cfm?id=4049&country=SRI%20LANKA accessed in February 2006

Jayaweera, D. (2001) 'Vehicle Inspection and Maintenance Policies and Programme – Sri Lanka', paper presented at the Global Initiative on Transport Emissions Asia Workshop 10–12 December 2001, Bangkok, Thailand

Senanayake, M. P., Samarakkody, R. P., Jasinghe, S. R., Hettiarachchi, A. P., Sumanasena, S. P. and Kudalugodaarachchi, J. (1999) 'Association Between Ambient Air Pollution and Acute Childhood Wheezy Episodes in Colombo', Forestry and Environment Symposium, Department of Forestry and Environmental Science, University of Sri Jayewardenpura, Sri Lanka, http://ybiol.tripod.com/forest/99sympo/9924sena.htm accessed in February 2006

World Bank (2003) 'Diesel Vehicles and Health Damage due to Air Pollution in Colombo', Washington DC, www.cleanairnet.org/caiasia/1412/article-35304.html accessed in February 2006

Yalegama, M. M. S. S. B. and Senanayake, N. (2004) 'Air Pollution and Contributions of Particulate Matter from Different Types of Diesel Vehicles in Sri Lanka', paper submitted to the Better Air Quality Workshop, Agra, India, 6–8 December 2004, www.cleanairnet.org/baq2004/1527/article-59285.html accessed in February 2006

Dhaka

Akhter, S., Islam M. A., Hossain S., Quadir, S. M. A., Khan, A., Begum, B., Khali-quzzaman, M. and Biswas, S. (2004) 'Air Quality Monitoring Program in Bangla-desh: Trends Analysis of Criteria Pollutants and Source Apportionment of Particulate Matter in Dhaka, Bangladesh', presented at the Better Air Quality Workshop in Agra, India, 6–8 December 2004, www.cleanairnet.org/baq2004/1527/article-59131.html accessed in February 2006

AQMP (2005) 'Ambient Air Quality Data Continuous Air Monitoring Station (CAMS)', Sangsad Bhaban, Dhaka, Bangladesh Department of Environment, www.doe-bd.org/ accessed November 2005

Biswas, S. K., Islam, A., Tarafdar S. A. and Khaliquzzaman, M. (2000) 'Monitoring of Atmospheric Particulate Matter (APM) in Bangladesh', paper presented at the Joint UNDP/RCA/IAEA Project Conference 'Sub Project: Air Pollution and its Trends', Manila, Philippines, 13–15 November 2000

Chowdhury, M. (2003) Personal Communication, Meshkat Chowdhury, Deputy Secretary, Energy and Mineral Resources Division, Ministry of Power, Energy and Mineral Resources, Government of Bangladesh

Core, J. (2003) 'An Action Plan for Dhaka, Bangladesh', paper presented at the Better Air Quality Workshop 2003, 17–19 December 2003, Edsa Shangri-La Hotel, Manila, Philippines, www.cleanairnet.org/baq2003/1496/article-58331.html accessed in February 2006

Dhaka City Website (2006) 'Dhaka City Corporation', www.dhakacity.org/html/about_dcc.html accessed March 2006

DOE (1997) 'The Environment Conservation Rules, 1997', Department of Environment, Dhaka, Bangladesh, www.doe-bd.org/2nd_part/179-226.pdf accessed in February 2006

DOE (2002) 'The Bangladesh Environment Conservation Act, 1995. Act No. 1', Unofficial English Version. Department of Environment, Dhaka, Bangladesh, www.doe-bd.org/2nd_part/153-166.pdf accessed in February 2006

DOE (2001) *Bangladesh State of Environment 2001*, Dhaka, Bangladesh, Department of Environment

DOE (2005a) 'About DOE', Department of Environment, www.doe-bd.org/core.html accessed in April 2006

DOE (2005b) 'Air Quality Management Project (AQMP)', Proposed Vehicle Emission Standards (VES), Dhaka, Bangladesh, Department of Environment, www.doe-bd.org/aqmp/standard.html accessed in April 2006

ICTP (2001) 'Urban Transport and Environment Improvement Study', ADB TA 3297-BAN, Final Report Summary, September 2001, Intercontinental Consultants and Technocrats Pvt. Ltd., India, www.cleanairnet.org/caiasia/1412/articles-36556_BAN_TA3297_Summary_Rpt.pdf accessed in February 2006

Islam, S. (2003) 'Dhaka gets AQI', *The Daily Star*, vol 4, p55, 20 July 2003, www.thedailystar.net/2003/07/20/d30720250278.htm accessed February 2006

Kaiser R., Henderson, A. K., Daleym, W. R., Naughton, M., Khan, M. H., Rahman, M., Kieszak, S. and Rubin, C. H. (2001) 'Blood Lead Levels of Primary School Children in Dhaka, Bangladesh', *Environmental Health Perspectives,* vol 109, pp563–566

World Bank (1998) *Air Pollution in Dhaka, Bangladesh*, Draft Report, Washington DC, The World Bank Group, South Asia Environmental Unit, 5 August 1998

Hanoi

ADB (2002) 'Integrated Action Plan to Reduce Vehicle Emissions in Vietnam', Asian Develoment Bank, Manila, www.adb.org/Vehicle-Emissions/actionviet.asp accessed in February 2006

Ha, L. (2004) 'Two Million Working Days Lost to Asthma', Vietnam Net Bridge, http://english.vietnamnet.vn/social/2004/11/347002/ accessed in February 2006

Hien, P. D., Bac, V. T., Lam, D. T. and Thinh, N. T. H. (2003) 'Sources of PM_{10} in Hanoi and Implications for Air Quality Management', paper presented at the Better Air Quality Workshop, 17–19 December 2003, Manila, www.cleanairnet.org/baq2003/1496/article-57919.html accessed in February 2006

Khaliquzzaman, M. (2005) 'Air Quality Issues in Hanoi', paper presented at Meeting on AQM for Hanoi Urban Transport Project, 26 January 2005, Vietnam, siteresources.worldbank.org/INTVIETNAM/Resources/MKZAirQualityIssuesHanoi4.pdf accessed in April 2006

Le, Tran Lam (2003) Personal Communication, Hanoi, DOSTE, 22 July 2003

Swisscontact (2005) 'Concepts for an Improved Air Quality Monitoring System and Emission Inventory for Hanoi', Hanoi, Vietnam, Swisscontact, www.cleanairnet.org/caiasia/1412/article-59548.html accessed in February 2006

Vietnam Panorama News Online (2005) 'Health Environment. Hanoi facing severe Air Pollution', *Vietnam Panorama News Online*, 30 November 2005, http://vietpan.com/categories/_e/5/a_detail.php?a_id=13704&lang=_e accessed in March 2006

Ho Chi Minh City

Dam, Q. T. (2004) 'A Vision for Cleaner Emissions from Motorcycles in Vietnam', presentation at the Cleaner Vehicles and Fuels in Vietnam Workshop, 13–14 May

2004, Hanoi, Vietnam, Vietnamese Ministry of Transportation and US-AEP, www.nea.gov.vn/Sukien_Noibat/Chatluong_KK/hoinghi-hoithao/Presentations%20ENGLISH/Dam%20Thang%20-%20Day%202.pdf accessed in April 2006

Dang, D. N. (1995) 'Pollution and Urban Areas: Reality and Solutions' ('O nhiem moi truong do thi: thuc trang va giai phap'), Saigon Giai Phong Thu Bay, so 276 (in Vietnamese) as quoted in Hiep Nguyen Duc (1996) *Some Aspects of Air Quality in Ho Chi Minh City, Vietnam*, proceedings of the Seminar on Environment and Development in Vietnam, National Centre for Development Studies, Australian National University, 6–7 December 1996, http://coombs.anu.edu.au/~vern/env_dev/papers/pap06.html accessed in February 2006

Department of Public Health (2001) *Annual Report*, Ho Chi Minh City, Vietnam, Department of Public Health

Do, H. O. (2003) 'Environmental Pollution Situation and Industrial Relocation Program in HCMC City', proceeding of the Second Meeting of the Kitakyushu Initiative Network, Weihai, China, 15–17 October, www.iges.or.jp/kitakyushu/Meetings/KIN2/Documents/Presentations%20and%20Papers%20(FROM%20MEETING)/15%20October/4%20Pilot%20Activities/City%20Presentations/Ho%20Chi%20Minh.pdf accessed in February 2006

HEPA (2006) Personal Communication, Than Dam, Ho Chi Minh Environmental Protection Agency, 17 February 2006

Sivertsen, B., Thanh, N., Khoa, L. V. and Dam, V. T. (2005) 'The Air Quality Monitoring and Management System for HCMC, Vietnam', Ho Chi Minh City Environmental Improvement Project, Air Quality Monitoring Component, Norwegian Institute for Air Research, Olso, Norway, www.nilu.no/niluweb/itemframe.cfm?lang=1&id=6864&type=2&senter=227 accessed in March 2006

Vietnam Register (2002) 'Integrated Action Plan to Reduce Vehicle Emissions in Vietnam', ADB, www.adb.org/Vehicle-Emissions/actionviet.asp accessed in April 2006

Hong Kong

Civic Exchange (2004) *Air Pollution: Air Quality Management Issues in the Hong Kong and the Pearl River Delta*, Civic Exchange White Paper, November 2004, Hong Kong, www.civic-exchange.org/publications/2004/airpollutionwhitepaper.pdf accessed 6 February 2006

EPD (2004a) 'Environmental Protection Department Homepage', www.epd.gov.hk/epd/english/environmentinhk/air/data/emission_inve.html accessed February 2006

EPD (2004b) 'Air Quality in Hong Kong 2003', Environmental Protection Department, Hong Kong, www.epd-asg.gov.hk/english/report/files/aqr03e.pdf accessed in February 2006

EPD (2005) 'Annual Air Quality Statistics 2005 (preliminary)', Environmental Protection Department, Hong Kong, www.epd-asg.gov.hk/english/report/files/aqr05e.pdf accessed in April 2006

Hong Kong (2003) 'The Environment: Air Pollution', www.info.gov.hk/yearbook/2003/english/chapter14/14_05.html accessed in February 2006

Mottershead, T. (2002) 'Hong Kong', in Mottershead, T. (ed.) *Environmental Law and Enforcement in the Asia-Pacific Rim*, Hong Kong, Singapore, Malaysia, Sweet and Maxwell Asia

Wong, C. M. (2002) 'Final Report for the Provision of Service for Study of Short Term Health Impact and Costs due to Road Traffic-related Air Pollution', Department of Community Medicine, the University of Hong Kong, Hong Kong SAR, www.epd. gov.hk/epd/english/environmentinhk/air/studyrpts/files/ap_health_impact_02.pdf accessed in February 2006

Wong, T. W., Lau, T. S., Yu, T. S., Neller, A., Wong, S. L., Tam, W. and Pang, S. W. (1999) 'Air Pollution and Hospital Admissions for Respiratory and Cardiovascular Diseases in Hong Kong', *Occupational and Environmental Medicine*, vol 56, pp679–683

Wong, C. M., Ma, S., Hedley, A. J. and Lam, T. H. (2001) 'Effect of Air Pollution on Daily Mortality in Hong Kong', *Environmental Health Perspectives*, vol 109, pp335–340

Wong, C. M., Atkinson, R. W., Anderson, H. R., Hedley, A. J., Ma, S., Chau, P. Y. and Lam, T. H. (2002) 'A Tale of Two Cities: Effects of Air Pollution on Hospital Admissions in Hong Kong and London compared', *Environmental Health Perspectives*, vol 110, pp67–77

Jakarta

ADB (2002) 'Action Plan: Integrated Vehicle Emission Reduction Strategy for Greater Jakarta, Indonesia', prepared by Indonesian Multi-Sectoral Action Plan Group on Vehicle Emissions Reduction for Asian Development Bank Regional Technical Assistance 5937, Reducing Vehicle Emissions in Asia, July 2002

Duki, Z., Sudarmadi, S., Suzuki, S. and Kawada, T. (2003) 'Effect of Air Pollution on Respiratory Health in Indonesia and its Economic Cost', *Archives of Environmental Health*, vol 58, no 3, pp135–43, www.ncbi.nlm.nih.gov/entrez/query.fcgi?cmd=R etrieve&db=PubMed&list_uids=14535572&dopt=Abstract accessed in February 2006

MoE (2005) Personal Communication, Amelia Rachmatunisa, Head, Air Quality Monitoring Sub Division, Ministry of Environment, Indonesia, May 2005

Nugroho, S. B. (2003) 'Lead monitoring for ambient air quality in Jabotabek areas', presentation at the Better Air Quality Workshop, 17–19 December 2003, Manila, Philippines, www.cleanairnet.org/baq2003/1496/articles-58110_resource_1.doc accessed in February 2006

Tamin, R. and Rachmatunisa, A. (2004) 'Integrated Air Quality Management in Indonesia', presented at the Better Air Quality Workshop, Agra, India, 6–8 December 2004, www.cleanairnet.org/baq2006/1757/articles-69951_sample.doc accessed in February 2006.

Wirahadikusumah, K. (2002) 'Jakarta Air Quality Management: Trends and Policies', presentation at the Regional Workshop on Better Air Quality in Asian and Pacific Rim Cities, 16–18 December 2002, Hong Kong, www.cse.polyu. edu.hk/~activi/BAQ2002/BAQ2002_files/Proceedings/Subworkshop1/sw1a-6Wirahadikusumah_paper.pdf accessed in February 2006

World Bank (1997) 'Urban Air Quality Management Strategy in Asia. Jakarta Report' World Bank Technical Paper No. 379, Washington DC, World Bank, www-wds. worldbank.org/servlet/WDS_IBank_Servlet?pcont=details&eid=000009265_3980313101856 accessed February 2006

World Bank (2003) 'World Bank Indonesia Environment Monitor 2003', Washington DC, World Bank, http://siteresources.worldbank.org/INTINDONESIA/Resources/Publication/03-Publication/indo_monitor.pdf accessed February 2006

Kathmandu

Business Age (2001) 'A Surgery of Nepalese Tourism', *Business Age*, vol 3, no 10, Kathmandu

CEN (2002) 'Bull's Trench Brick Kilns to be Banned in Kathmandu', *Clean Energy News*, vol 2, p15, 12 March 2002, www.environmentnepal.com.np/news_d.asp?id=45 accessed in April 2006

CEN (2003) 'Health Impacts of Kathmandu's Air Pollution', Factsheet 6, Clean Energy Nepal, Kathmandu, Nepal, www.keva.org.np/publication/KEVA's%20Publication/CEN%20Fact%20Sheet%206%20-%20Health_Eng.pdf accessed in July 2006

CEN/ENPHO (2003) 'Health Impacts of Kathmandu's Air Pollution', Kathmandu, Clean Energy Nepal/Environment and Public Health Organization, www.dec.org/pdf_docs/PNACW355.pdf accessed in February 2006

ENPHO (2001) 'Status of Brick Kilns in the Kathmandu Valley', Kathmandu, Environment and Public Health Organization

Gautam, C., Sharma, S. and Fuglsang, K. (2004) 'Air Quality Monitoring and Management in Kathmandu, Nepal', paper presented at the Better Air Quality Workshop,

6–8 December 2004, Agra, India, www.cleanairnet.org/baq2004/1527/article-59132.html accessed in February 2006

KFW (2004) 'Nepal: Rehabilitation of Chobhar Cement Factory', ('Nepal: Rehabilitierung der Zementfabrik Chobhar Abschlusskontroll- und Schlussprüfung Projekt Rehabilitierung der Zementfabrik Chobhar', in German), Kreditanstalt fuer Wiederaufbau – Enwicklungsbank, www.kfw-entwicklungsbank.de/DE_Home/Evaluierung/Weitere_Informationen/Schlussprf90/PDF-Dokumente/nepal_chobhar.pdf accessed in March 2006

LEADERS Nepal (1998) 'A Citizen Report on Air Pollution in Kathmandu Valley: Children's Health at Risk. Leaders Nepal, Kathmandu. Nepal', quoted in *UNEP State of Environment, Nepal 2001*, United Nations Environment Programme, Bangkok, www.rrcap.unep.org/reports/soe/nepal_air.pdf accessed in February 2006

MOEST (2005) 'Ambient Air Quality of Kathmandu Valley 2003–2004', Ministry of Environment, Science and Technology, Singh Durbar, Kathmandu

MOPE (2005) 'Air Pollution Data of Kathmandu Valley (PM_{10})', Ministry of Population and Environment (MOPE), www.mope.gov.np accessed in February 2006

Sapkota, B. K., Sharma, N. P., Poudel, K. and Bhattarai, B. (1997) 'Particulate Pollution Levels in Kathmandu Valley', *Environment*, vol 3, no 2, Kathmandu, Ministry of Population and Environment

Shresta, R. M. and Raut, A. (2002) 'Air Quality Management in Kathmandu', presentation at Better Air Quality in Asian and Pacific Rim Cities (BAQ 2002) Workshop, 16–18 December 2002, Hong Kong, www.cse.polyu.edu.hk/~activi/BAQ2002/BAQ2002_files/Proceedings/CityFocus/cf-6Shrestha_paper.pdf accessed in February 2006

World Bank (1997) *Urban Air Quality Management Strategy in Asia (URBAIR) Kathmandu Valley Report*, World Bank, Washington DC

Kolkata

Bhattacharya, P. (2005) Personal Communication, Scientist, West Bengal Pollution Control Board

Chakraborti, D. (2003) 'Kolkata City: An Urban Air Pollution Perspective', in Whitelegg, J. and Haq, G. (eds) *World Transport Policy and Practice*, London, Earthscan

CMDA (2005) 'Kolkata Metropolitan Development Authority', www.cmdaonline.com/ accessed in February 2006.

CSE (2006) *The Leapfrog Factor: Cleaning Air in Asian Cities*, New Delhi, India, Centre for Science and Environment

Lahiri, T., Roy, S., Basu, C., Ganguly, S., Ray, M. R. and Lahiri, P. (2000a) 'Air Pollution in Calcutta Elicits Adverse Pulmonary Reaction in Children', *Indian Journal of Medical Research*, vol 112, pp21–26

Lahiri, T., Ray, M. R., Mukherjee, S., Basu, C. and Lahiri, P. (2000b) 'Marked Increase in Sputum Alveolar Macrophages in Residents in Calcutta: Possible Exposure Effect of Severe Air Pollution', *Current Science*, vol 78, pp399–404, www.ias. ac.in/currsci/feb252000/RESEARCHARTICLE.pdf accessed in February 2006

MOEF (2002) 'The Air (Prevention and Control of Pollution) Act, 1981' Ministry of Environment and Forests, http://envfor.nic.in:80/legis/air/air1.html, accessed in February 2006

Roy, S., Ray, M. R., Basu, C., Lahiri, P. and Lahiri, T. (2001) 'Abundance of Sidero-phages in Sputum. Indicator of an Adverse Lung Reaction to Air Pollution', *Acta Cytologica*, vol 45, pp958–964

WBDoE (2003) 'Air Quality. Ambient Air Quality and Sources of Pollution', Department of Environment, Government of West Bengal

WBPCB (2003) *A Quinquenniel Report April 1998 – March 2003*, Kolkata, India, West Bengal Pollution Control Board

World Bank (2004) *Toward Cleaner Urban Air in South Asia: Tackling Transport Pollution, Understanding Sources*, Washington DC, Joint UNDP/World Bank Energy Sector Management Assistance Programme (ESMAP)

Metro Manila

Amador, J. (2003) 'Developments in Air Quality Monitoring in Metro Manila', presentation at the Better Air Quality Workshop, Manila, Philippines, 17–19 December 2003, www.cleanairnet.org/baq2003/1496/article-57774.html accessed in February 2006

Anglo, E. (2004) 'Modeling the Dispersion of Particulates in Metro Manila', presented at Sweeping the Dust off Manila's Air: A Science and Policy Dialogue 21 April 2004, www.observatory.ph/aq/presentations/07_Anglo_Modeling/index. html accessed in February 2006

Briones, M. D. D. (2002) 'Effects of Air Pollution on the Spirometric Findings of Filipino School Children: A Challenge to the Clean Air Act', Philippine Children's Medical Center, www.cleanairnet.org/caiasia/1412/article-69622.html accessed in March 2006

CAA (1999) Clean Air Act of 1999 (Republic Act 8749), Philippines, www.elaw. org/assets/pdf/philippines.caa.pdf accessed in February 2006

DENR (2005) *National Air Quality Status Report (2003-2004)*, Department of Environment and Natural Resources, Manila (draft, for publication)

LGC (1991) 'The Local Government Code of the Philippines. Republic Act No. 7160', Chan Robles Virtual Law Library, www.chanrobles.com/localgov.htm accessed in February 2006

LTO (2005) 'Plans and Programs. Private Emission Testing Centers', Land Transportation Office, www.lto.gov.ph/programs.html accessed in April 2006

Santiago, R. (2003) 'Roadside Enforcement of MV Emission Standard: Metro Manila Experience', presented at the ENVIROTECH 2003 Regional Workshop and Exhibit, 8 October 2003, Jakarta, Indonesia

Torres, E., Ronald, D. Subida, R. D., Gapas, J. L., Sarol, J. N., Villarin, J. T.,Vinluan, R. J. N., Ramos, B. M. and Quirit, L. L. (2004) *Public Health Monitoring: A Study Under the Metro Manila Air Quality Improvement Sector Development Program (MMAQISDP)*, Manila, Asian Development Bank

Vergel, K. (2004) 'Urban Transportation and the Environment in Metro Manila: Programs and Initiatives for Clean Air', Department of Built Environment Seminar, Tokyo Institute of Technology, 2 April 2000, www.cv.titech.ac.jp/~jsps/research_report/K.Vergel.pdf accessed in June 2006

Villarin, J. T., Anglo, E. G., Simpas, J. B., Choy, R. N., Cheng Chua, K. U., Lorenzo, G. H. and Uy, S. N. Y. (2004) 'Urban Air Quality Group, The Manila Observatory. Climate Studies Division', presentation at Clearing the Air: A Forum on Monitoring and Reporting Air Quality in Metro Manila, 5 October 2004, www.cleanairnet.org/caiasia/1412/articles-59026_villarin.ppt accessed in February 2006

Walsh, M. P. (2000) 'The Motor Vehicle Pollution Control Program of the Philippines', *Carlines*, vol 4, pp22–26

World Bank (2001) 'Project Appraisal Document for the Metro Manila Urban Transport Integration Project', www.gefweb.org/Documents/Project_Proposals_for_Endorsem/Philippines_Marikina_Bikeways.pdf accessed in February 2006

Mumbai

Kamat, S. R. (2000) 'Mumbai Studies of Urban Air Pollution and Health Resulting Synergistic Effects of SPM, SO_2 and NO_x', International Conference on Environmental and Occupational Respiratory Diseases, Indian Toxicological Research Centre, Lucknow, India

MCGM (2004) 'Municipal Corporation of Greater Mumbai Environment Status Report 2002–2004', www.mcgm.gov.in accessed in February 2006

NEERI (2005) Personal Communication, Anjali Srivastava, Senior Assistant Director, National Environmental Engineering Research Institute, Mumbai Zonal Centre, Mumbai, India

Parikh, J. K. and Hadker, N. (2003) 'Economic Impacts of Urban Air Pollution: Valuation for Mumbai, India', *International Journal of Environment and Pollution*, vol 19, pp498–515

Shankar, P. R. and Rao, G. R. (2002) 'Impact of Air Quality on Human Health: A Case of Mumbai City, India', paper presented at the IUSSP Regional Conference on Southeast Asia's Population in a Changing Asian Context, 10–13 June 2002, Bangkok, Thailand, www.iussp.org/Bangkok2002/S09Shankar.pdf accessed in February 2006

Singh, R. N., (2002), 'Air Quality Management in Indian Cities: Trends and Challenges', presentation at Better Air Quality in Asian and Pacific Rim Cities (BAQ 2002), Hong Kong Convention and Exhibition Centre, Hong Kong

Srivastava, A. and Kumar, R. (2002) 'Economic Valuation of Health Impacts of Air Pollution in Mumbai', *Environmental Monitoring and Assessment*, vol 75, pp135–143

World Bank (1997) 'Urban Air Quality Management Strategy in Asia: Greater Mumbai Report', World Bank technical paper No. 381, Washington DC, World Bank

New Delhi

Agarwal, K. S., Mughal, M. Z., Upadhyay, P., Berry, J. L. and Puliyel, J. M. (2002) 'The Impact of Atmospheric Pollution on Vitamin D Status of Infants and Toddlers in Delhi, India', *Archives of Disease in Childhood*, vol 87, pp111–113, http://adc.bmjjournals.com/cgi/content/full/87/2/111 accessed in April 2006

CPCB (1998) *Status of Air Quality in National Capital Region*, New Delhi, India, Central Pollution Control Board, Ministry of Environment and Forests, Government of India

CPCB (2001a) *Air Quality in Delhi 1989–2000*, New Delhi, India, Central Pollution Control Board, Ministry of Environment and Forests

CPCB (2001b) 'Central Pollution Control Board Annual Report for 2000–2001', Ministry of Environment and Forests, Government of India, New Delhi, India, www.cpcb.nic.in/annual.htm accessed in February 2006

CPCB (2003a) *National Ambient Air Quality Status 2001*, New Delhi, India, Central Pollution Control Board, Ministry of Environment and Forests, p3

CPCB (2003b) 'Parivesh Newsletter: June 2003', R&D for Pollution Control, New Delhi, India, Central Pollution Control Board Initiatives, Ministry of Environment and Forests

CPCB (2005a) 'Environmental Standards: Approved Fuels in the National Capital Territory of Delhi', Central Pollution Control Board, www.cpcb.nic.in/standard67.htm accessed in April 2006

CPCB (2005b) *National Ambient Air Quality Status, National Ambient Air Quality Monitoring Series, NAAQMS/26/2004–05*, New Delhi, India, Central Pollution Control Board, Ministry of Environment and Forests, December, p3

CPCB (2005c) 'National Ambient Air Quality Status 2003, National Ambient Air Quality Monitoring Series', NAAQMS/26/2004–05, Central Pollution Control Board, New Delhi, www.cpcb.nic.in/ accessed in April 2006

Chhabra, S. K., Chhabra, P., Rajpal, S. and Gupta, R. K. (2001) 'Ambient Air Pollution and Chronic Respiratory Morbidity in Delhi', *Archives of Environmental Health*, vol 56, pp58–64

Cropper, M. L., Simon, N. B., Alberini, A. and Sharma, P. K. (1997) *The Health Effects of Air Pollution in Delhi*, India, PRD Working Paper No. 1860, www.worldbank.org/nipr/work_paper/1860/ accessed in February 2006

Duggal, V. K. and Pandey, G. K. (2002) 'Air Quality Management in Delhi', presentation at BAQ 2002, Hong Kong, 16–18 December 2002, www.cse.polyu.edu hk/~activi/BAQ2002/presentations.htm accessed in February 2006

Dursbeck, F., Erlandsson, L. and Weaver C. (2001) 'Status of Implementation of CNG as a Fuel for Urban Buses in Delhi', New Delhi, India, Centre for Science and Environment, 23 May 2001

FADA (2005) 'Demystifying RTO Queries – Emission Norms', Federation of Automobile Dealers Associations, New Delhi, www.fadaweb.com/drq_emission.htm accessed in April 2006

GONCT (2003) *Cleaner, Greener, Naturally: Welcome to World's First CNG City*, New Delhi, India, Government of National Capital Territory of Delhi

Gurjar, B. R., Lelieveld, J. and Van Aardenne, J. (2003) 'Air Quality and Emission Trends in a Megacity: The Case of Delhi', in R. S. Sokhi and J. Brechler (eds) *Proceedings of the 4th International Conference on Urban Air Quality: Measurement, Modelling and Management*, Prague, Czech Republic, Charles University, 25–27 March 2003

Mashelkar, R. A., Biswas, D. K., Mathur, O. P., Natarajan, R., Nyati, K. P., Singh, D. V. and Singhal, S. (2002) Report of the Expert Committee on Auto Fuel Policy. Government of India, New Delhi, http://petroleum.nic.in/ accessed in February 2006

MoEF (1983) 'The Air (Prevention and Control of Pollution) Act, 1981', Ministry of Environment and Forests, New Delhi, India, http://envfor.nic.in:80/legis/air/air3.html accessed in February 2006

MoEF (1997) 'White Paper on Air Pollution in Delhi with an Action Plan', Ministry of Environment and Forests, New Delhi, India, http://envfor.nic.in/divisions/cpoll/delpolln.html accessed in February 2006

Seoul

AFMA/KEI (2002) *International Joint Research Project on Air Pollution in the Megacities of Asia*, Seoul, Republic of Korea, Ministry of Environment of Korea, Korean Environment Institute

Kwon, H. J., Cho, S. H., Chun, Y., Lagarde, F. and Pershagen, G. (2002) 'Effects of the Asian Dust Events on Daily Mortality in Seoul, Korea', *Environmental*

Research Section A, vol 90, pp1–5, http://yellow.metri.re.kr/study/study_e_2002_01.pdf accessed in April 2006

Kwon, H. J. and Cho S. H. (1999) 'Air Pollution and Daily Mortality in Seoul', *Korean Journal of Preventative Medicine*, vol 32, pp36–52

Kwon, H. J., Cho, S. H., Nybergh, F. and Pershagen, G. (2001) 'Effects of Ambient Air Pollution on Daily Mortality in a Cohort of Patients with Congestive Heart Failure', *Epidemiology*, vol 12, pp413–419

Lee, J. T. and Schwartz, J. (1999) 'Reanalysis of the Effects of Air Pollution and Daily Mortality in Seoul, Korea: A Case-crossover Design', *Environmental Health Perspectives*, vol 107, pp633–636

Lee, J. T., Shin, D. and Hung, Y. (1999) 'Air Pollution and Daily Mortality in Seoul and Ulsan, Korea', *Environmental Health Perspectives*, vol 107, pp149–154

Lee, J. T., Kim, H., Hong, Y.-C., Kwon, H.-J., Schwartz, J. and Christiani, D.C. (2000) 'Air Pollution and Daily Mortality in Seven Major Cities of Korea, 1991–1997', *Environmental Research A*, vol 84, pp247–254

MoL (2002) 'Ministry of Legislation', Seoul, www.moleg.go.kr accessed in February 2006

MOE (2002) *Annual Report of Air Quality in Korea 2001*, Ministry of Environment, Seoul, Republic of Korea, http://eng.me.go.kr/user/ accessed in February 2006

MOE (2003) *Annual Report of Air Quality in Korea 2002*, Ministry of Environment, Seoul, Republic of Korea, http://eng.me.go.kr/user/ accessed in February 2006

Shanghai

Fu, J. (2004) Personal Communication, Shanghai Environmental Monitoring Centre, Shanghai, 19 June 2004

Jingguang, L. (2003) 'An Update in Efforts to Promote Cleaner Vehicles in China', *The Sinosphere Journal*, vol 6, no 1, pp28–32

Kan, H. and Chen, B. (2003a) 'A Case-crossover Study of Ambient Air Pollution and Daily Mortality in Shanghai', *Journal of Occupational Health*, vol 45, pp119–124

Kan, H. and Chen, B. (2003b) 'Air Pollution and Daily Mortality in Shanghai: A Time-series Study', *Archives of Environmental Health*, vol 58, no 6, pp360–367

Kan, H., Jia, J. and Chen, B. (2004a) 'A Time-series Study on the Association of Stroke Mortality and Air Pollution in Zhabei District, Shanghai', *Wei Sheng Yan Jiu*, vol 33, no 1, pp36–38

Kan, H., Jia, J. and Chen, B. (2004b) 'The Association of Daily Diabetes Mortality and Outdoor Air Pollution in Shanghai, China', *Journal of Environmental Health*, vol 67, no 3, pp21–26

SEPA (2003) 'Environmental Standards in China', State Environmental Protection Administration, www.zhb.gov.cn/english/ accessed in April 2006

UNESCAP (2000) 'Integrating Environmental Considerations into the Economic Decision-Making Process. Modalities for Environmental Assessment: Urban Development and Environmental Protection in Shanghai', United Nations Economic and Social Commission for Asia and the Pacific, www.unescap.org/drpad/publication/integra/modalities/china/4ch000ct.htm accessed in February 2006

Wang, J. (2003) 'Retrospection and Perspective of Air Pollution Control in Shanghai', paper presented at the 3rd Thematic Seminar: Urban Air Quality Management, Bangkok, Thailand, 20–21 February 2003

Singapore

Chew, F. T., Goh, D. Y. T. and Lee, B. W. (1999a) 'Under-recognition of Childhood Asthma in Singapore: Evidence from a Questionnaire Survey', *Annals of Tropical Paediatrics*, vol 19, pp83–91

Chew, F. T., Goh, D. Y., Ooi, B. C., Saharom, R., Hui, J. K. and Lee, B. W. (1999b) 'Association of Ambient Air-pollution Levels with Acute Asthma Exacerbation among Children in Singapore', *Allergy*, vol 54, no 4, pp320–329

Emmanuel, S. C. (2000) 'Impact to Lung Health of Haze from Forest Fires: The Singapore Experience', *Respirology*, vol 5, no 2, pp175–182

Goh, D. Y. T., Chew, F. T., Quek, S. C. and Lee, B. W. (1996) 'Prevalence and Severity of Asthma, Rhinitis and Eczema in Singapore Schoolchildren', *Archives of Disease in Childhood*, vol 74, pp131–135

NEA (2004a) 'A Code of Practice on Pollution Control (with amendments 2001, 2002 and 2004)', National Environment Agency, Singapore, www.nea.gov.sg/cms/pcd/coppc_2002.pdf accessed in February 2006

NEA (2004b) Personal Communication, Soh Suat Hoon, National Environment Agency, Singapore, 28 October 2004

NEA (2005) 'Laws Administered by the National Environmental Agency', National Environment Agency Singapore, http://app.nea.gov.sg/cms/htdocs/category_sub.asp?cid=180 accessed in February 2006

Quah, E. and Boon, T. L. (2003) 'The Economic Cost of Particulate Air Pollution on Health in Singapore', *Journal of Asian Economics*, vol 14, no 1, pp73–90

Surabaya

ADB (2002) 'Action Plan: Integrated Vehicle Emission Reduction Strategy for Greater Jakarta, Indonesia', Asian Development Bank Regional Technical Assistance 5937:

Reducing Vehicle Emissions in Asia, Manila, Philippines, www.adb.org/vehicle-emissions/actionindo.asp accessed in February 2006

Air Laboratory of Surabaya (2003) *Report of Operation Ambient Air Quality Monitoring in 2003*, Surabaya, Indonesia, Environment Department of Surabaya

GEF (1998) 'Research Report on Trends in Environmental Considerations Related to Overseas Activities of Japanese Companies FY 1997', Global Environment Forum, www.env.go.jp/earth/coop/oemjc/ind/e/contents.html accessed in March 2006

Silaban, T. (2003) 'Ambient Air Quality Management in Surabaya, Indonesia', presented at the 3rd Thematic Seminar: Kitakyushu Initiative Seminar on Urban Air Quality Management, 20–21 February 2003, United Nations Conference Centre, Bangkok, Thailand, www.unescap.org/esd/environment/kitakyushu/urban_air/city_report/Surabaya2.pdf accessed in February 2006

Taipei

BEP (2003) *Air Quality Management Plan 2003*, Taipei, Bureau of Environmental Protection

Chang, C. C. S. S., Ho, S. C. and Yang, C. Y. (2005) 'Air Pollution and Hospital Admissions for Cardiovascular Disease in Taipei, Taiwan', *Environmental Research*, vol 98, no 1, pp114–119

Chen, P. C., Lai, Y. M., Wang, J. D., Yang, C. Y., Hwang, J. S., Kuo, H. W., Huang, S. L. and Chan, C. C. (1998) 'Adverse Effect of Air Pollution on Respiratory Health of Primary School Children in Taiwan', *Environmental Health Perspectives*, vol 106, no 6, pp331–335

Chen, P. C., Lai, Y. M., Chan, C. C., Hwang, J. S., Yang, C. Y. and Wang, J. D. (1999) 'Short-Term Effect of Ozone on the Pulmonary Function of Children in Primary School', *Environmental Health Perspectives*, vol 107, no 11, pp921–925

Chen, Y. S., Sheen, P. C., Chen, E. R., Liu, Y., K. Wuc, T. N. and Yang, C. Y. (2004) 'Effects of Asian Dust Storm Events on Daily Mortality in Taipei, Taiwan', *Environmental Research*, vol 95, no 2, pp151–155

Chuang, J. S. C. (2001) 'Taiwan Motorcycle Emission Control Experience', paper presented at the Reduction of Emissions from 2–3 Wheelers Workshop, Hanoi, Vietnam, 5–7 September

DEP (2003) 'Environmental Monitoring and Analysis', Taipei City, Department of Environmental Protection, http://163.29.36.51/dep/index.jsp?recordid=2208 accessed in February 2006

DEP (2004) 'Air Quality of Taipei City in 2003', Taipei City, Department of Environmental Protection, http://163.29.36.51/dep/index.jsp?recordid=2232 accessed in February 2006

DUD (2003) 'An Overview: Taipei City Today', Department of Urban Development, www.planning.taipei.gov.tw/TCDB/OverView_Today.asp accessed in February 2006

EPA (2002) 'Air Pollution Control Act of 1975. Amended 2002', Environmental Protection Administration, Executive Yuan, Republic of China, http://law.epa.gov.tw/en/laws/889404502.html accessed in February 2006

EPA (2003a) 'Environmental Management – Environmental Quality Monitoring: Data Assurance', Environmental Protection Administration, www.epa.gov.tw/english/webezA-6/code/main2.asp?catNo=4&subcatNo=4&cat=Env.%20Quality%20Monitoring accessed in March 2006

EPA (2003b) 'Toxic Chemical Sustances Management Act', Environmental Protection Administration, http://law.epa.gov.tw/en/laws/808150657.html accessed in March 2006

EPA (2003c) 'Environmental Impact Assessment Act', Environmental Protection Administration, http://law.epa.gov.tw/en/laws/379692190.html accessed in March 2006

EPA (2003d) 'Air Quality and Forecast Report', Environmental Protection Administration, www.epa.gov.tw/english/webezA-5/code/main1.asp?catNo=3&cat=Air%20Pollution%20Control accessed in April 2006

Guo, Y, L., Lin, Y. C., Sung, F. C., Huang, S. L., Ko, Y. C., Lai, J. S., Su, H. J., Shaw, C. K., Lin, R. S. and Dockery, D. W. (1999) 'Climate, Traffic-Related Air Pollutants, and Asthma Prevalence in Middle-School Children in Taiwan', *Environmental Health Perspectives,* vol 107, no 12, pp1001–1006

Hwang, J. S., Chen, Y. J., Wang, J. D., Lai, Y. M., Yang, C. Y. and Chan, C. C. (2000) 'Subject-Domain Approach to the Study of Air Pollution Effects on Schoolchildren's Illness Absence', *American Journal of Epidemiology,* vol 152, no 1, pp67–74

Yang, C. Y., Chang, C. C., Chuang, H. Y., Tsai, S. S., Wu, T. N. and Ho, C. K. (2004) 'Relationship between Air Pollution and Daily Mortality in a Subtropical City: Taipei, Taiwan', *Environment International*, vol 30, no 4, pp519–523

Tokyo

BoE (2005) 'The Environment in Tokyo 2005', Bureau of Environment, Tokyo, Japan, www2.kankyo.metro.tokyo.jp/kouhou/env/eng/pdf/all.pdf accessed in February 2006

Dieselnet (2005a) 'Emission Standards – Japan: On-road Vehicles and Engines', Dieselnet, www.dieselnet.com/standards/jp/onroad.html accessed in April 2006

Dieselnet (2005b) 'Emission Standards – Japan: Automotive NOx and PM Law: Law Concerning Special Measures to Reduce the Total Amount of Nitrogen Oxides and

Particulate Matter Emitted from Motor Vehicles in Specified Areas', Dieselnet, www.dieselnet.com/standards/jp/noxpmlaw.html accessed in April 2006

MOE (2005a) 'Laws and Regulations', Tokyo, Ministry of the Environment, www. env.go.jp/en/lar/index.html accessed in February 2006

MOE (2005b) 'Environmental quality standards in Japan – Air quality', Ministry of the Environment, Tokyo, www.env.go.jp/en/air/aq/aq.html accessed April 2006

Piver, W. T., Ando, M., Ye, F. and Portier, C. J. (1999) 'Temperature and Air Pollution as Risk Factors for Heat Stroke in Tokyo, July and August 1980–1995', *Environmental Health Perspectives*, vol 107, pp911–916

Sekine, K., Shima, M., Nitta, Y., Adachi, M. (2004) 'Long Term Effects of Exposure to Automobile Exhaust on the Pulmonary Function of Female Adults in Tokyo, Japan', *Occupational and Environmental Medicine*, vol 61, pp350–357

TMG (2004) 'Principal Policies of the Tokyo Metropolitan Government. Restrictions on Diesel Vehicles', Tokyo Metropolitan Government, Tokyo, www.chijihon.metro. tokyo.jp/english/PROFILE/POLICY/policy6.htm accessed in April 2006

Voorhees, A. S., Araki, S., Sakai, R. and Sato, H. (2000) 'An Ex Post Cost–Benefit Analysis of the Nitrogen Dioxide Air Pollution Control Program in Tokyo', *Journal of the Air Waste Management Association*, vol 50, pp391–410

Ye, F., Piver, W. T., Ando, M. and Portier, C. J. (2001) 'Effects of Temperature and Air Pollutants on Cardiovascular and Respiratory Diseases for Males and Females Older than 65 Years of Age in Tokyo, July and August 1980–1995', *Environmental Health Perspectives*, vol 109, pp355–359

four

Development of Air Quality Management in Asian Cities

INTRODUCTION

The emergence of urban air pollution problems in Asia and the ability to manage deteriorating air quality is influenced by the level and speed of economic development. These can have a direct effect on the severity of environmental pollution and the strategies and capacity to address these issues.

The results of the air quality management (AQM) capability survey presented in Chapter 2 and the information on air quality levels collated in Chapter 3 can be used to examine the link between air quality and AQM capability in the 20 cities.

This chapter discusses the key aspects of AQM. It attempts to identify common characteristics and barriers to improving air quality and outlines the necessary measures that each group of cities might take to achieve better air quality.

EFFECTIVE AIR QUALITY MANAGEMENT

The principles on which the management of air quality is based on ensuring the protection of human health and environment from air pollution. These principles include the right to clean air for all, access to environmental information and an awareness of the air pollution situation. AQM is based on the polluter pays, protection and precautionary principles and ensures a cost-effective approach using best available technology is taken. Yet a number of economic, institutional and political constraints may hamper the full implementation of all these guiding principles (APMA/CAI-ASIA, 2004).

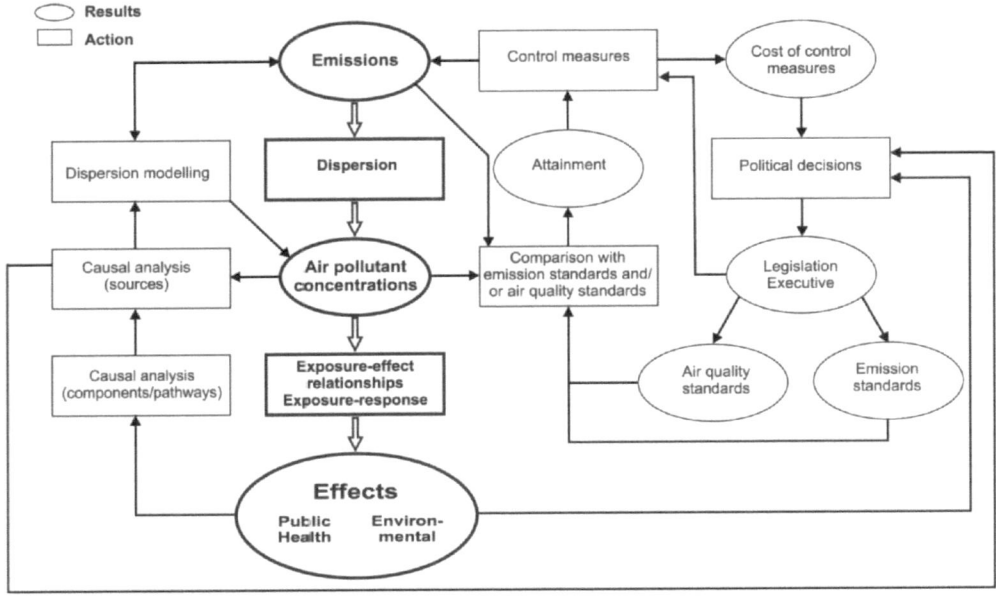

Figure 4.1 *A simplified framework for air quality management*

Source: APMA/CAI-ASIA (2004)

Figure 4.1 presents a simplified framework for AQM. The effectiveness of an AQM strategy is dependent on the implementation of a number of measures. These include comprehensive legislation, emission inventories, air quality monitoring, dispersion modelling, exposure and damage assessments and emission standards. However, legislative powers and resources to implement and enforce air pollution regulation and the availability of a range of cost-effective pollution control measures are vital to improving air quality.

Some form of air quality monitoring is undertaken in the majority of Asian cities. However, the extent of monitoring, the types of pollutants monitored and the reliability of data collated vary widely between cities and countries in the region often reflecting the level of economic development. At present monitoring in peri-urban and rural areas is rarely undertaken. Yet this is essential in order to understand the potential impacts on agriculture and ecosystems and the contribution of long-range pollution to urban air pollution.

Air quality standards are another important aspect of AQM. In areas where air quality standards are exceeded on a regular basis, measures need to be taken to reduce air pollution levels. National air quality standards, however, are often based on a number of constraints such as economic and technological feasibility, which do not guarantee that they are protecting human health and the environment. The WHO air quality guidelines (WHO, 2000; 2005) are derived from epidemiological and toxicological studies in such a way as to minimize the risk of health and environmental impacts. For PM, SO_2 and O_3, however, it has been generally accepted that existing studies provide no indication of any reliable threshold of effect.

National and local government authorities in Asia have adopted a range of air quality standards either based on WHO guidelines or USEPA standards. These have tended to differ from country to country (see Tables 4.1–4.6). However, the majority of Asian countries have more lenient standards than the guidelines or limit values prescribed by WHO or the EU. In particular, standards for PM_{10} have been largely based on USEPA limits. With a move in EU countries to a 50 µg/m^3 limit for 24-hour averages of PM_{10} and 40 µg/m^3 as an annual mean and the setting of guideline values by WHO for PM_{10} and $PM_{2.5}$, there is a need to review current PM standards in Asia (EU, 1999; WHO, 2005).

COMPARATIVE AIR QUALITY

The AQM capability of the 20 cities can be compared to measured levels of PM_{10}, NO_2 and SO_2 in the period 2000–2004 and the extent to which EU limits and WHO guidelines are exceeded.

Table 4.7 presents the guideline values used for the categorization of air quality in the 20 Asian cities. EU air quality limits for PM_{10} and NO_2 are used since these limits are the lowest values which are considered to be technologically achievable in urban areas. The EU limit values are associated with the lowest risk to public health caused by exposure to air pollutants. The values are based on most recent health effects studies and are more stringent than USEPA standards. With respect to SO_2, the 2000 WHO SO_2 guideline value is used. This is because the 2005 WHO guidelines with a value of 20 µg/m^2 for 24-hour exposure do not provide an annual mean guideline value. Such a guideline value in principle, would be well below 20 µg/m^3. The annual EU limit value (20 µg/m^3) was not used because it protects ecosystems.

The levels of O_3 are not considered here because nine cities (Beijing, Colombo, Dhaka, Hanoi, Kolkata, Metro Manila, Mumbai, Shanghai and Tokyo) did not report sufficient O_3 data and the data quality for four cities (Ho Chi Minh City, Jakarta, Kathmandu and Surabaya) is unknown.

Due to the difference in air quality monitoring, reporting procedures, quality of data, source and year of data only a subjective assessment of air quality in the 20 cities can be given. This assessment is based on the most recent data collated and

Table 4.1 *Particulate matter standards*

Compound Country	Cities	TSP [µg/m³]					PM₁₀ [µg/m³]			PM₂.₅ [µg/m³]	
		1 hr	3 hrs	8 hrs	24 hrs	1 yr	1 hr	24 hrs	1 yr	24 hrs	1 yr
WHO								50	20	25	10
EU								50	40		
USEPA								150	50	65	15
Bangladesh								150	50	65	15
China (Class II)					300	200		150	100		
Hong Kong, SAR, China					260	80		180	55		
India (residential, rural & other areas)					200	140		100	60		
Indonesia					230	90		150			
	Jakarta				230	90		150			
	Surabaya				230	90		150			
Japan							200	100			
Republic of Korea								150	70		
	Busan							150	70		
	Seoul							120	60		
Nepal					230			120			
Philippines					230	90		150	60		
Singapore								150	50	65	15
Sri Lanka		500	450	350	300	100					
Taiwan, China								150	65		
Thailand					330	100		120	50		
Vietnam								150	50		

Table 4.2 *Sulphur dioxide standards*

Compound		SO$_2$ [µg/m³]				
Country	Cities	10 min	1hr	8 hrs	24 hrs	1yr
WHO		500			20	
EU			350		125	20
USEPA					365	78
Bangladesh					365	80
China (Class II)			500		150	60
Hong Kong, SAR, China			800		350	80
India (residential, rural & other areas)					80	60
Indonesia			900		365	60
	Jakarta		900		260	60
	Surabaya		900		365	60
Japan			266		105	
Republic of Korea*						
	Busan		392		131	57
	Seoul		314		105	26
Nepal					70	50
Philippines					180	80
Singapore					365	78
Sri Lanka			200	120	80	
Taiwan					140	79
Thailand*			785		314	105
Vietnam					125	50

Note: * Conversion factor for ppb to µg/m³: 2.616.

Table 4.3 *Nitrogen dioxide standards*

Compound	NO$_2$ [µg/m^3]				
Country	Cities	1 hr	8 hrs	24 hrs	1 yr
WHO		200			40
EU		200			40
USEPA					100
Bangladesh					100
China (Class II)		240		120	80
Hong Kong, SAR, China		300		150	80
India (residential, rural & other areas)				80	60
Indonesia		400		. 150	100
	Jakarta	400		92.5	60
	Surabaya	400		150	100
Japan		75 to 113			
Republic of Korea*		282		150	94
	Busan	282		150	94
	Seoul	263		132	75
Nepal				80	40
Philippines				150	
Singapore					100
Sri Lanka		250	150	100	
Taiwan					95
Thailand*		325			
Vietnam		200			40

Note: * Conversion factor for ppb to µg/m^3: 1.880.

Table 4.4 *Ozone standards*

Compound	O₃ [µg/m³]				
Country	Cities	1 hr	8 hrs	24 hrs	1 yr
WHO			100		
EU			120		
USEPA		235			
Bangladesh		235	157		
China (Class II)		160			
Hong Kong, SAR, China		240			
India (residential, rural & other areas)					
Indonesia		235			50
	Jakarta	200			30
	Surabaya	235			50
Japan		118			
Republic of Korea*		200	120		
	Busan	200	120		
	Seoul	200	120		
Nepal					
Philippines		140	60		
Singapore		235			
Sri Lanka		200			
Taiwan			120		
Thailand*		200			
Vietnam		120		80	

Note: * Conversion factor for ppb to µg/m³: 1.962.

Table 4.5 *Carbon monoxide standards*

Compound	CO [µg/m³]					
Country	Cities	15 min	30 min	1hr	8 hrs	24 hrs
WHO		100,000	60,000	30,000	10,000	
EU					10,000	
USEPA				40,000	10,305	
Bangladesh				40,000	10,000	
China	(Class II)			10,000		4000
Hong Kong, SAR, China				30,000	10,000	
India (residential, rural & other areas)				4000	2000	
Indonesia				30,000		10,000
	Jakarta			26,000		9000
	Surabaya			30,000		10,000
Japan					22,900	11,450
Republic of Korea*				28,625	10305	
	Busan			28,625	10,305	
	Seoul			28,625	10,305	
Nepal		100,000			10,000	
Philippines				35,000	10,000	
Singapore				40,000	10,305	
Sri Lanka				30,000	10,000	
Taiwan					10,000	
Thailand*				34,350	10,305	
Vietnam				30,000	10,000	

Note: * Conversion factor for ppb to µg/m³: 1.145.

Table 4.6 *Lead standards*

Compound		Pb [µg/m³]				
Country	Cities	1 hr	24 hrs	1 month	3 months	1 yr
WHO						0.5
EU						0.5
USEPA					1.5	
Bangladesh						0.5
China (Class II)						
Hong Kong, SAR, China					1.5	
India (residential, rural & other areas)			1			0.75
Indonesia						1.0
	Jakarta					1.0
	Surabaya					1.0
Japan						
Republic of Korea						
	Busan					0.5
	Seoul				1.0	0.5
Nepal						0.5
Philippines					1.5	1.0
Sri Lanka			2			0.5
Taiwan, China				1		
Thailand				1.5		
Vietnam			5			

Table 4.7 *Guideline values used for categorization of air quality in 20 Asian cities*

Air pollutant	Guideline value / standard ($\mu g/m^3$)	Exposure time	Interpretation	Source
PM_{10}	40	1 year	Average of all concentrations monitored during a year	EU (1999)
NO_2	40	1 year		
SO_2	50	1 year		WHO (2000)

available to the authors. Figure 4.2 presents an overview of the relative air quality situation in each city. Three broad categories of air pollution can be determined:

1 **Serious pollution:** Annual EU and WHO values are exceeded by more than a factor of two for annual concentrations of PM_{10}, NO_2 and SO_2.
2 **Moderate pollution:** EU and WHO values are exceeded by up to a factor of two for annual concentrations of NO_2 and SO_2. Annual concentrations of PM_{10} are above half the EU limit (20 $\mu g/m^3$) and exceed this limit by up to a factor of two (80 $\mu g/m^3$).
3 **Low pollution:** Annual concentrations of NO_2 and SO_2 meet WHO guidelines or are lower. Annual concentrations of PM_{10} are less than half the EU limit (20 $\mu g/m^3$).

Table 4.8 shows that in 13 cities health impacts studies have been undertaken. PM_{10} and TSP were considered as being responsible for health impacts in ten cities (Bangkok, Beijing, Hong Kong, Mumbai, New Delhi, Seoul, Shanghai, Singapore, Taipei and Tokyo) of which six have moderate and three (Beijing, New Delhi and Shanghai) serious PM pollution (see Figure 4.2). NO_2 was found as a probable causal agent in Hong Kong, Mumbai, Seoul, Shanghai, Singapore, Taipei and Tokyo, which have low to moderate pollution. Health impacts attributed to SO_2 were found in Beijing, Hong Kong, Mumbai and Shanghai, which have low to moderate pollution. This illustrates that concentrations of SO_2 below the 2000 WHO guideline value may have health effects, which is in line with the recent reduction in the guideline value (WHO, 2005).

In order to determine if there have been any changes in the urban air quality situation, the four categories can be used to re-evaluate previous assessments of

City	PM$_{10}$	NO$_2$	SO$_2$
Bangkok	◉	◉	○
Beijing	●	⊗	◉
Busan	◉	◉	○
Colombo	◉	◉	◉
Dhaka	●	◉	○
Hanoi	●	○	○
Ho Chi Minh City	◉	○	○
Hong Kong	◉	◉	○
Jakarta	●	◉	◉ ⊘
Kathmandu	●	⊗	⊗
Kolkata	●	◉	○
Metro Manila	◉	⊗	⊗
Mumbai	◉	○	○
New Delhi	●	○	○
Seoul	◉	◉	○
Shanghai	●	◉	◉
Singapore	◉	○	○
Surabaya	◉	○	◉
Taipei	◉	◉	○
Tokyo	◉	◉	○

Legend ● Serious　◉ Moderate　○ Low　⊗ No data　⊘ Data of unknown quality

Figure 4.2 *Overview of air quality in 20 Asian cities*

air quality in 14 Asian cities undertaken by UNEP/WHO (1992) and WHO/UNEP/ MARC (1996) (see Figure 4.3). In order to make the findings comparable for PM it has been assumed that PM$_{10}$ is equal to half of TSP.

A comparison of Figures 4.2 and 4.3 shows an improvement (categorization from a more polluted to a less polluted class) in the air quality situation in Bangkok, Metro Manila and Mumbai for PM$_{10}$ concentrations. There has also been an improvement in Beijing and Seoul with regard to levels of SO$_2$ and New Delhi with regard to NO$_2$. Deterioration in air quality (categorization from a less polluted to a more polluted class) is observed for Busan, Jakarta and Taipei with regard to NO$_2$ and Jakarta with regard to SO$_2$. Where there has been no change in categorization for a particular pollutant, this does not mean that improvements have not been achieved in lowering air pollutant concentrations but rather that the improvement was not sufficient to change the category.

On a pollutant-by-pollutant basis the following situation is given in the 20 Asian cities.

Table 4.8 *Overview of health studies undertaken in 20 Asian cities*

City	Health studies	Compounds associated with observed health impacts
Bangkok	Y	PM_{10}
Beijing	Y	TSP, SO_2
Busan	Y	SO_2
Colombo	N	–
Dhaka	N	–
Hanoi	N	–
Ho Chi Minh City	N	–
Hong Kong	Y	NO_2, O_3, SO_2, PM_{10}
Jakarta	N	–
Kathmandu	N	–
Kolkata	Y	'air pollution'
Metro Manila	Y	'air pollution'
Mumbai	Y	SO_2, NO_2, TSP
New Delhi	Y	TSP, Pb
Seoul	Y	NO_2, O_3, PM_{10}
Shanghai	Y	PM_{10}, SO_2, NO_2
Singapore	Y	PM_{10}, TSP, O_3
Surabaya	N	–
Taipei	Y	PM_{10}, NO_2, CO, O_3
Tokyo	Y	NO_2, PM_{10}

Note: Y (N) = health studies were (not) performed.

Particulate matter

PM_{10} continues to be a problem in all cities although concentrations have decreased in at least three cities of the 14 Asian cities that were considered in the UNEP/WHO (1992) and WHO/UNEP/MARC (1996) assessments. Figure 4.2 shows that eight cities (Beijing, Dhaka, Hanoi, Jakarta, Kathmandu, Kolkata, New Delhi and Shanghai) are

City	PM$_{10}$	NO$_2$	SO$_2$
Bangkok	●	○	○
Beijing	●	◓	●
Busan	○	○	○
Hong Kong	○	◓	○
Jakarta	●	○	○
Kolkata	●	⊗	◓
Metro Manila	●	⊗	○
Mumbai	●	⊗	○
New Delhi	●	◓	◓
Seoul	○	○	●
Shanghai	●	◓	◓
Singapore	○	◓	○
Taipei	○	○	○
Tokyo	○	◓	○
Legend ● Serious		◓ Moderate ○ Low ⊗ No data	

Figure 4.3 *Overview of air quality in 14 Asian cities based on the assessments of WHO/UNEP (1992) and WHO/UNEP/MARC (1996)*

categorized as having serious PM$_{10}$ pollution. Health impacts due to exposure to PM$_{10}$ such as increased daily mortality and morbidity due to cardio-respiratory conditions have been reported in only a few of these cities classified as seriously polluted (Beijing, Kolkata, New Delhi and Shanghai). Twelve cities (Bangkok, Busan, Colombo, Ho Chi Minh City, Hong Kong, Metro Manila, Mumbai, Seoul, Singapore, Surabaya, Taipei and Tokyo) have moderate PM$_{10}$ pollution. Health impacts due to PM pollution have been reported in all cities except Busan, Colombo and Surabaya (e.g. Cropper et al, 1997; Chew et al, 1999; Ye et al, 2001; Wong et al, 2002; Chang et al, 2003, 2005, Tamura et al, 2003; Torres et al, 2004;   Kan et al, 2004a). In Kolkata and Metro Manila (both of which have serious and moderate PM levels respectively) 'air pollution' was identified as responsible agent for observed health impacts (Shankar and Rao, 2002 and Lahiri, 2004).

Nitrogen dioxide

None of the cities has NO$_2$ levels categorized as serious. Eleven cities have moderate NO$_2$ pollution (Bangkok, Busan, Colombo, Dhaka, Hong Kong, Jakarta, Kolkata, Seoul, Shanghai, Taipei and Tokyo). In three cities (Busan, Jakarta and Taipei) NO$_2$ concentrations have increased from low to moderate levels for the period 1992 to 2004 compared with previous assessments (UNEP/WHO, 1992; WHO/UNEP/MARC, 1996). Health impacts possibly associated with NO_2 exposure reported in Hong Kong include hospital admissions for all respiratory diseases, all cardiovascular diseases, chronic obstructive pulmonary diseases and heart failure, and admissions for asthma, pneumonia and influenza (Wong et al, 1999, 2001, 2002). In Seoul, exposure to NO_2 in combination with O_3 has been related to health impacts such as respiratory illness (Infante-Rivard, 1993) and exacerbation of allergic asthmatic responses (Lewis et al, 2000). In Shanghai, a significant increase in the relative risk of non-accident mortality, of cardiovascular diseases and of chronic obstructive pulmonary diseases was associated with a 10 µg/m³ increase in NO_2 (Kan and Chen, 2003a, 2003b; Kan et al, 2004a). A similar increase of NO_2 levels was found to be correlated with a significantly increased relative risk of diabetes mortality in Shanghai (Kan et al, 2004b). In Tokyo, a study demonstrated a strong association between concentrations of NO_2, and emergency cases with myocardial infarction, acute bronchitis and cardiac insufficiency (Ye et al, 2001). In addition, NO_2 was a significant risk factor for all age groups and gender (Piver et al, 1999). Another study suggested that long-term exposure to NO_2 was related to an increase in airway resistance and a chronic decrease in pulmonary lung function.

Figure 4.2 shows that six cities (Hanoi, Ho Chi Minh City, Mumbai, New Delhi, Singapore and Surabaya) comply with the EU limit value. The risk for health effects due to NO_2 exposure in these cities is, therefore, low. However, impacts on health due to NO_2 have been observed in Mumbai (Kamat, 2000).

Sulphur dioxide

Thirteen cities comply with the annual SO_2 air quality guideline value and are categorized as having low SO_2 pollution (see Figure 4.2). Only Beijing, Colombo, Jakarta, Shanghai and Surabaya do not comply with the WHO guideline value and are categorized as having moderate SO_2 pollution. Compared with the 1992 and 1996 assessments, SO_2 pollution in Beijing, New Delhi and Seoul has decreased while in Jakarta it has increased. However, SO_2 levels below the WHO guideline value of 50 µg/m³ can still have adverse health effects (Xu et al, 1994). For example, in Shanghai a significant increase in the relative risk of non-accident mortality of cardiovascular diseases and of chronic obstructive pulmonary diseases was associated with a 10 µg/m³ increase in SO_2 in the presence of PM_{10} and NO_2 (Kan and Chen, 2003a; 2003b; Kan et al, 2004a). A similar increase of SO_2 levels was found to correlate with a significantly increased relative risk of diabetes mortality in Shanghai (Kan et al, 2004b). Analysis of the most recent epidemiological studies from developed countries, mostly Europe, has led WHO to reduce the 24-hour guideline value from 125 µg/m³ to 20 µg/m³ and not to determine an annual guideline value.

STATUS OF AQM

In order to better distinguish between the cities within the same AQM capability category as determined in Chapter 2, the original capability scoring can be further refined. Table 4.9 shows this distinction with four cities deemed Excellent (I) and three Excellent (II). Those cities with lower scores can learn from higher scoring cities on how they can improve. This is also true of course for cities in the lower AQM categories.

Table 4.9 *Qualitative classification of air quality management capabilities*

Original capability scoring	Original capability classification	New capability scoring	New capability classification	Cities
81–100	Excellent	91–100	Excellent I	Hong Kong, Singapore, Taipei, Tokyo
		81–90	Excellent II	Bangkok, Seoul, Shanghai
61–80	Good	71–80	Good I	Beijing, Busan
		61–70	Good II	New Delhi
41–60	Moderate	51–60	Moderate I	Ho Chi Minh City, Jakarta, Kolkata, Metro Manila, Mumbai
		41–50	Moderate II	Colombo
21–40	Limited	31–40	Limited I	Hanoi, Surabaya
		21–30	Limited II	Dhaka, Kathmandu
0–20	Minimal	0–20	Minimal	—

Figure 4.4 combines the AQM capabilities categories of the 20 cities shown in Table 4.10 and the comparative assessment of air quality in each city as described above. It also indicates the tendencies of air pollutant concentrations as more or less decreasing ⬇, more or less increasing ⬆ and essentially constant ➡. The indication of air quality tendencies is based on air quality information collected and covers a period of 5–10 years. This information is summarized in the city profiles in Chapter 3 and is indicative of the action the city is currently taking in air pollution abatement or plans to take in the future.

Figure 4.4 shows that an excellent AQM capability does not always equate to low air pollutant concentrations. For example, Bangkok and Hong Kong have excellent AQM capability but have moderate NO_2 pollution while other cities with lower AQM capability have lower NO_2 pollution. Another example is Shanghai, which while disposing of excellent capability still has serious air pollution.

Cities with serious PM_{10} pollution tend to have a lower AQM capability (good, moderate and limited). For example, three out of four cities (Dhaka, Hanoi and Kathmandu) with a limited AQM capability have serious PM_{10} pollution. Cities with an excellent AQM capability have taken successful action to address PM_{10} pollution.

Capability classification	City	PM_{10}	NO_2	SO_2
Excellent I	Hong Kong	Moderate, No change	Moderate, No change	Low, No change
	Singapore	Moderate, No change	Low, No change	Low, Down
	Taipei	Moderate, Down	Moderate, Down	Low, Down
	Tokyo	Moderate, Down	Moderate, Down	Low, Down
Excellent II	Bangkok	Moderate, No change	Moderate, Up	Low, Down
	Seoul	Moderate, No change	Moderate, Up	Low, No change
	Shanghai	Serious, No change	Moderate, No change	Moderate, Down
Good I	Beijing	Serious, Down	No data	Moderate, Down
	Busan	Moderate, Down	Moderate, No change	Low, Down
Good II	New Delhi	Serious, Down	Low, Up	Low, Down
Moderate I	Ho Chi Minh City	Moderate, No change	Low	Low
	Jakarta	Serious, Up	Moderate, Up	Moderate, Data of unknown quality
	Kolkata	Serious, Down	Moderate, Up	Low, Down
	Metro Manila	Moderate, Down	No data	No data
	Mumbai	Moderate, No change	Low, Down	Low, Down
Moderate II	Colombo	Moderate, No change	Moderate, Down	Low, Down
Limited I	Hanoi	Serious, Down	Low, No change	Low, Up
	Surabaya	Moderate, Data of unknown quality	Low, Data of unknown quality	Low, Data of unknown quality
Limited II	Dhaka	Serious, Data of unknown quality	Moderate, Data of unknown quality	Low, Data of unknown quality
	Kathmandu	Serious, Data of unknown quality	No data	No data

Legend ● Serious ◍ Moderate ○ Low ⊗ No data
Up No change Down ⓘ Data of unknown quality

Figure 4.4 *Assessment of air quality management capabilities and trends*

Eight cities (Beijing, Busan, Hanoi, Kolkata, Metro Manila, New Delhi, Taipei and Tokyo) show a downward tendency for PM_{10} concentrations, only Jakarta shows an upward tendency. Eight cities (Bangkok, Colombo, Ho Chi Minh City, Hong Kong, Mumbai, Seoul, Singapore and Shanghai) show a constant tendency of PM_{10} concentrations. For three cities a tendency cannot be determined due to the unknown quality of the data. The constant tendency for PM_{10} concentrations in Hong Kong, and Singapore (Excellent I) and Bangkok and Seoul (Excellent II) indicate the difficulty in reducing particle concentrations in these cities. The upward tendency of PM_{10} concentrations in Jakarta (Moderate I) requires urgent attention. It should be noted that Singapore and Tokyo are at the lower end of the moderate class, while most other cities with PM_{10} concentrations classified moderate are at the upper end (see Chapter 3).

Five cities (Bangkok, New Delhi, Jakarta, Kolkata and Seoul) exhibit an upward tendency for NO_2 pollution. Five cities exhibit a constant tendency (Busan, Hanoi, Hong Kong, Shanghai, Singapore). Of these cities, Busan (Good I), Hong Kong (Excellent I) and Bangkok and Shanghai (Excellent II) have moderate NO_2 pollution. Four cities (Colombo, Mumbai, Taipei and Tokyo) show a downward tendency of NO_2 concentrations.

In ten cities (Bangkok, Beijing, Busan, Colombo, Kolkata, Mumbai, New Delhi, Shanghai, Taipei and Tokyo) there is a downward tendency in SO_2 concentrations. All these cities have been categorized as having low SO_2 pollution with the exception of Beijing and Shanghai. In Hong Kong and Seoul the tendency is constant and the concentrations comply with the WHO guideline value, indicating that a baseline concentration has been achieved. Hanoi is the only city with an increasing tendency of SO_2 pollution, which, however, still appears to be low.

Although the 20 Asian cities considered here have underlying similarities in their air pollution problems, many differences also exist. Cities with excellent or good AQM capabilities still experience serious or moderate PM_{10} and NO_2 pollution (see Table 4.10).

CITY SPECIFIC OBSERVATIONS

Dhaka, Kathmandu (Limited II)

Dhaka and Kathmandu have been classified as having Limited (II) AQM capacity and both have serious PM_{10} pollution. The moderate NO_2 levels in Dhaka are mostly caused by motor vehicle pollution. Integrated measures to address the import of old cars from other more developed countries, better fuel quality, and implementation and enforcement of inspection and maintenance programmes would assist in reducing

Table 4.10 *PM$_{10}$ and NO$_2$ level categorization versus capability scoring*

PM pollution	City	Capability classification
Serious	Shanghai	Excellent
	Beijing, New Delhi	Good
	Jakarta, Kolkata	Moderate
	Dhaka, Hanoi, Kathmandu	Limited
Moderate	Bangkok, Hong Kong, Seoul, Singapore, Taipei, Tokyo	Excellent
	Busan	Good
	Colombo, Ho Chi Minh City, Metro Manila, Mumbai	Moderate
	Surabaya	Limited
NO$_2$ pollution		
Moderate	Hong Kong, Bangkok, Seoul, Taipei, Tokyo, Shanghai	Excellent
	Busan	Good
	Colombo, Jakarta, Kolkata	Moderate
	Dhaka	Limited
Low	Singapore,	Excellent
	New Delhi	Good
	Ho Chi Minh City, Mumbai	Moderate
	Hanoi, Surabaya	Limited

this problem. The monitoring capability of Kathmandu is presently being limited to PM and the data are of unknown quality. Due to steadily increasing motorization, NO$_2$, CO and HC emissions may substantially increase in the future. With increasing NO$_2$ and HC, O$_3$ concentrations may also become be a future problem.

In order to have effective AQM it is necessary to improve the monitoring capabilities in both cities, including the long-term monitoring, kerbside and spatially representative monitoring of NO$_2$, CO, SO$_2$, O$_3$ and HC. For this task a hybrid network consisting of a few automatic and many diffusive samplers would be appropriate. Mapping techniques should be used for the spatial representation of the data.

Since air pollution episodes are not forecasted and no warnings given to the population, the capacity of dispersion modelling and use of models for forecasting should be enhanced.

Hanoi and Surabaya (Limited I)

Hanoi and Surabaya have been classified as having Limited (I) AQM capacity. In spite of improvements, the monitoring capacity of these cities is still limited to short-term monitoring for all relevant air pollutants. The data do not yet allow tendencies or trends to be determined. The data for Surabaya are of unknown quality. The spatial distribution of pollutants cannot yet be assessed and kerbside monitoring is still absent. As a consequence, no use of mapping techniques is made. Major components of AQM are missing: dispersion modelling and pollutant episode forecasting is not possible due to lack of an emissions inventory. Exposure and epidemiological studies are non-existent.

The absence of key components of AQM such as reliable emission inventories makes understanding the causes of air pollution difficult. Actions taken are ad hoc in nature and cannot solve the problem of moderate to serious PM_{10} levels in the two cities.

Colombo (Moderate II)

Colombo has been classified as having Moderate (I) AQM capacity. The monitoring capacity of Colombo could be enhanced to include O_3 monitoring. Due to the limited number of monitoring stations, the spatial distribution of pollutants cannot yet be assessed. As a consequence no use of mapping techniques is made. Key components of AQM such as dispersion modelling and pollutant episode forecasting are not applied in Colombo due to the lack of an emissions inventory. The development of emission inventories would assist in providing a better understanding of the sources of the moderate PM_{10} levels and in developing measures to be taken as actions to mitigate pollution. Exposure and epidemiological studies are also non-existent in Colombo.

Ho Chi Minh City, Jakarta, Kolkata, Metro Manila and Mumbai (Moderate I)

Ho Chi Minh City, Jakarta, Kolkata, Metro Manila and Mumbai have been classified as having Moderate (I) AQM capability. The capacity of these cities to monitor all key pollutants is still limited. While PM_{10} is monitored in all cities, gaseous compounds, in particular CO and O_3, are not always addressed. Mapping techniques are rarely applied even if the spatial distribution of air pollutants is addressed. Emissions inventories do not adequately address domestic/commercial and other (e.g. aviation and shipping) sources. Dispersion modelling and pollutant episode forecasting is rarely applied. Exposure and epidemiological studies, if available, are rarely used as inputs for policy.

The moderate PM_{10} levels in Ho Chi Minh City, Metro Manila and Mumbai and the serious PM_{10} levels in Jakarta and Kolkata are probably due to local and transboundary sources. However, the contribution from long-range sources is unknown. An assessment of the influence of long-range sources and development of regional concepts would improve the situation. The moderate NO_2 levels in Jakarta and Kolkata are mainly due to local motor vehicle pollution.

New Delhi (Good II)

New Delhi has been classified as having Good (II) AQM capacity. The monitoring network in the city is fairly well developed. However, long-term monitoring of O_3 is not yet in place which prevents trends from being determined. In spite of good spatial coverage of monitoring sites, mapping techniques are not applied. Emissions inventories do not cover domestic, commercial and other sources (e.g. aviation and shipping). Various attempts have been made to estimate emissions of industrial, vehicle and power plant sources; however, the results have been somewhat contradictory (Mashelkar et al, 2002). The influence of long-range transport of pollution has not been assessed. There is no episode forecasting in New Delhi and no exposure assessment, while only a few epidemiological studies exist.

Serious PM_{10} concentrations and moderate NO_2 concentrations continue in New Delhi mostly due to local emissions. A better characterization of local and long-range sources will assist in formulating policies for effective AQM.

Beijing and Busan (Good I)

Beijing and Busan have been classified as having Good (I) AQM capacity. Established monitoring networks exist in both cities. Weaknesses in AQM include the lack of sophisticated statistical techniques in the evaluation of data, the completeness of emissions inventories, pollution forecasting, rare exposure assessment and scarcity of epidemiological studies.

In spite of the good capability of AQM in Beijing, PM_{10} concentrations are still classified as serious, while no data to determine NO_2 levels are available (data are only available for NO_x). In contrast, both PM_{10} and NO_2 levels in Busan are moderate. In both cities local levels are affected by long-range transport of air pollutants, which requires an integrated approach to AQM.

Bangkok, Seoul, Shanghai (Excellent II) and Hong Kong, Singapore, Taipei and Tokyo (Excellent I)

Bangkok, Seoul and Shanghai have been classified as having Excellent (II) AQM capacity. PM_{10} concentrations are serious in Shanghai and moderate in Bangkok and Seoul. NO_2 levels are also moderate in the three cities. In Shanghai SO_2 levels are also above the WHO guideline. Hong Kong, Singapore, Taipei and Tokyo have Excellent (I) AQM capacity. In spite of this, all cities have moderate PM_{10} concentrations while three cities have moderate NO_2 levels. Singapore is the only city with low NO_2 levels. The observations for the seven cities indicate that excellent capacity as defined in the capability questionnaire survey does not mean that concentrations are below acceptable levels. The increasing number of motor vehicles and kilometres travelled contribute to the high NO_2 and PM_{10} concentrations which have offset reductions in emissions gained from vehicle technology. In addition, transboundary pollution from mainland China has affected PM_{10} levels in Tokyo, Taipei, Seoul and Hong Kong while air quality in Singapore has been affected by transboundary influences from Indonesia (Koe et al, 2001; Lam et al, 2005; Reuther, 2000; Wang et al, 2003).

AQM AND ECONOMIC STATUS

The overall AQM capability index score of the 20 Asian cities can be used to consider the relationship between the wealth of a city as measured by the purchasing power parity (PPP) and AQM capability. The PPP, calculated from gross national income (GNI) in US$, is considered to have greater international comparability than the GNP (PRD, 2005). The PPP is an alternative exchange rate such that it is representative of a basket of goods in differing countries if the currencies are exchanged at that rate.

Figure 4.5 shows on a logarithmic scale a linear relationship between the cities, scores and the PPP/capita of the countries. A linear relationship can be discerned indicating that the higher the city score, the higher is the PPP/capita. However, Figure 4.6 shows that the linear increase in AQM capability with increasing PPP/capita is only true in the low PPP/capita range, that is at a PPP/capita below approximately US$7000 at which an AQM capability score of more than 80 can be achieved. A further increase in PPP/capita can only marginally increase the AQM capability of a city as it is already close to its maximum value of 100.

The overall AQM capability, therefore, does not increase significantly with the wealth of a country. Once the wealth of a country exceeds approximately US$7000 PPP/capita, expenditure on AQM is no longer dependent on the availability of resources. This observation was also made in the WHO/UNEP/MARC (1996) assessment. As a consequence other policy measures take on a more prominent role in

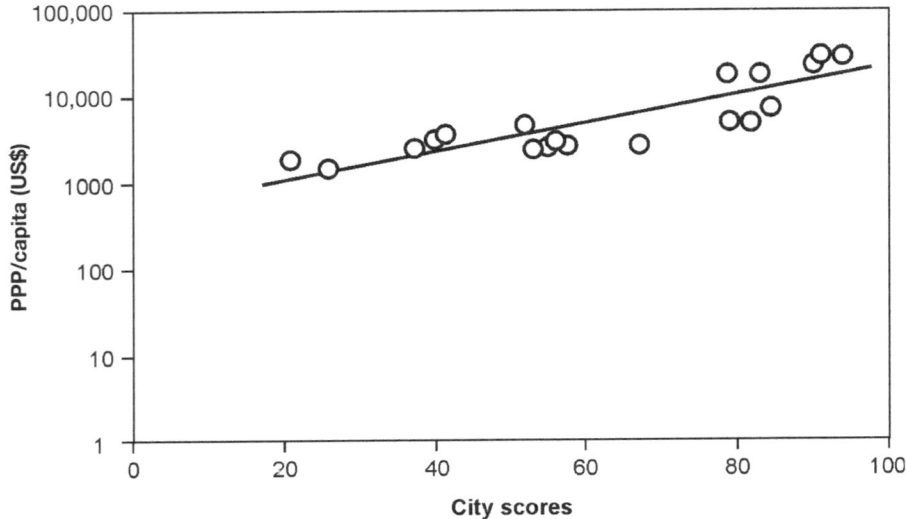

Figure 4.5 *Linear relationship between city scores and PPP/capita on a logarithmic scale*

Source: PRD (2005)

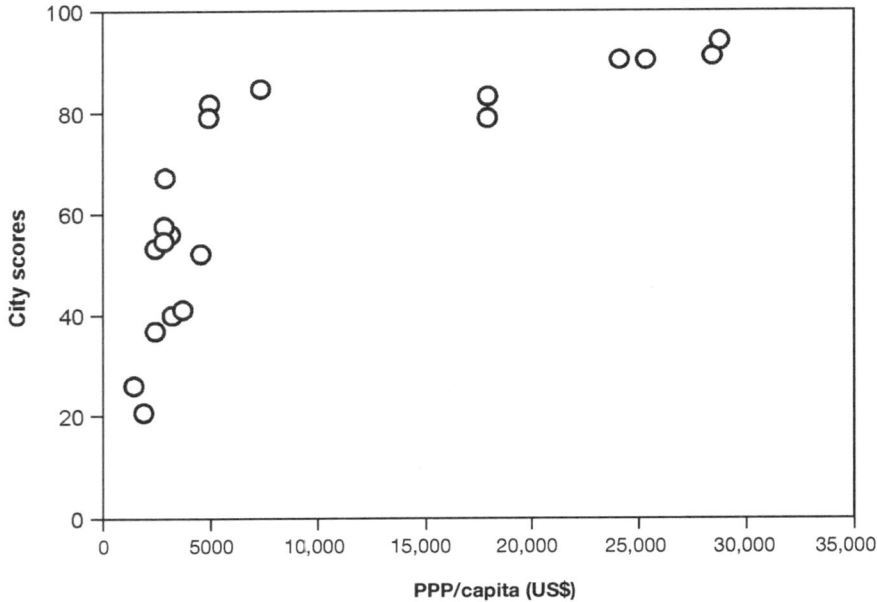

Figure 4.6 *The relationship between PPP/capita and city scores*

improving existing air quality. These include better implementation and enforcement of existing legislation and the introduction of measures to address transboundary air pollution. Considering that some of the cities which have an excellent to good AQM capability also have serious to moderate NO_2 and PM_{10} pollution, then further action is needed to address the issue of motor vehicle emissions and, if relevant, industrial and other sources.

For motor vehicles this may need to take the form of measures to reduce the demand to travel, by improving public transport and urban planning, promoting non-motorized transport and making cities safer and more inviting for walking and cycling (Haq, 1997; Whitelegg and Haq, 2003). In order to assess the PM_{10} contribution of industries, source apportionment and emission inventories would assist in identifying the major contributors and consequently determine the appropriate actions that need to be taken.

AIR QUALITY TRENDS AND DEVELOPMENT

In Chapter 1 the general development model of air pollution problems in cities according to their development status was described (UNEP/WHO, 1992). Using this model, air pollution exposure and the development of AQM capacity of 20 Asian cities, five general stages of AQM development can be determined. Table 4.11 describes each stage where the levels of air pollution concentrations associated with each stage are considered to be dependent on how early or late emission controls are initiated.

Figure 4.7 presents the different stages of AQM development according to air pollutant concentrations. Depending on which stage the city is at in the development of its air pollution problems, each key element of AQM will have a different priority and importance for the city. For some air pollutants such as O_3 and CO, air quality monitoring in a given city may still be rudimentary and require further expansion. Other cities will have well-developed monitoring systems but require the development of emission inventories and assessment of health impacts.

Many cities in Europe and North America have followed this development path and are currently experiencing relatively low levels of air pollution. For example, the City of London (UK), in 1952, experienced the London smog that lasted for five days and, according to research by Bell and Davis (2001), led to approximately 12,000 more deaths than usual. This smog episode resulted in the first UK Clean Air Act in 1956. The Act was aimed at controlling domestic sources of smoke pollution by introducing smokeless zones. Within ten years of the Act being adopted smoke emissions from industry were reduced by 74 per cent with those from domestic sources becoming the main polluter (Clapp, 1994). The introduction of cleaner coals led to a reduction in SO_2 pollution and the move to the use of natural gas reduced domestic emissions. The

Table 4.11 *AQM capacity, economic and air quality development*

Stage	AQM capacity	Level of economic development	Air quality development	Cities
I	Minimal	Increased urbanization, industrialization and motorization. Ad hoc AQM action applied	Deterioration of air quality through rising levels of air pollution	None
II	Limited	Urbanization, industrialization and motorization continues. Initial systematic AQM procedures applied	High but stabilizing levels of air pollution. Serious health and environmental impacts	Dhaka, Hanoi, Kathmandu, Surabaya
III	Moderate	Cleaner processes developed. Systematic AQM procedures developed	Air pollution decreasing from high levels	Colombo, Ho Chi Minh City, Jakarta, Kolkata, Metro Manila, Mumbai
IV	Good	Maturing of cleaner processes and use of cleaner fuels. Mature emission controls	Further improvement of air pollution	Beijing, Busan, New Delhi
V	Excellent	High technology applied	Low air pollution	Bangkok, Hong Kong, Seoul, Singapore, Shanghai, Taipei, Tokyo

City's air quality has improved considerably over the past fifty years progressively moving through each stage of development from Stages I to V. Nowadays, London is currently addressing the problem of vehicle-related air pollution and photochemical smog. Between 12 and 15 December 1991, London experienced the most severe NO_2 pollution since regular monitoring began in 1971, one-hour levels of NO_2 at 809 $\mu g/m^3$, greatly exceeding the WHO air quality guideline value of 200 $\mu g/m^3$. Levels of benzene increased by six to ten times in typical value. The 1991 London smog episode is claimed to have caused 160 extra deaths (Elsom, 1996).

While the model shown in Figure 4.7 may be true for compounds such as SO_2 and CO it differs for others such as PM_{10} and NO_2. This may be due to:

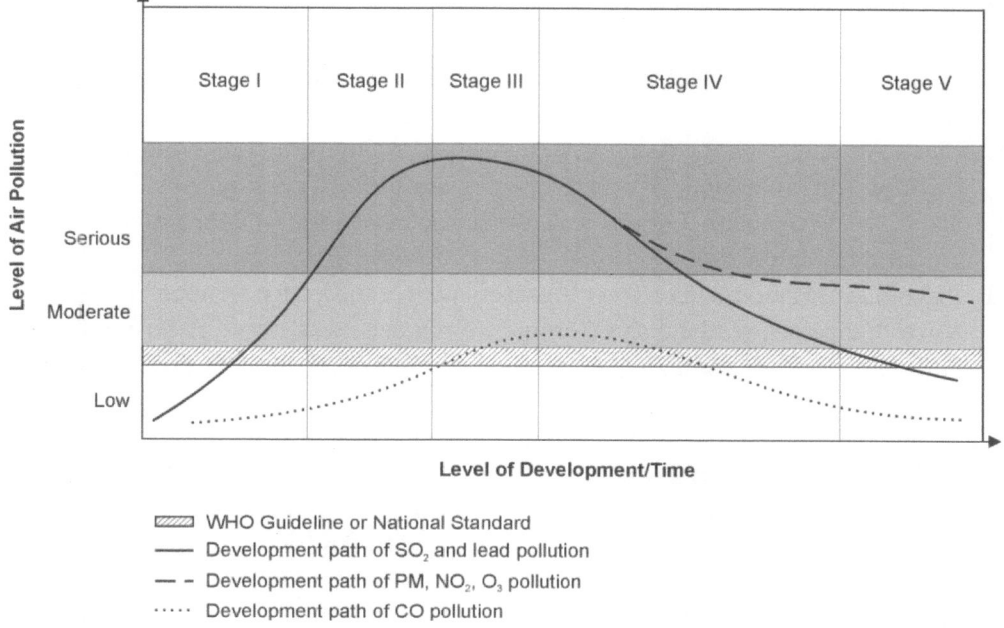

Figure 4.7 *Development path of air pollution problems in cities*

- gaps in implementation and enforcement of existing legislation;
- annual air quality standards for PM in Asian cities being less stringent than USEPA standards, EU limit values and WHO guideline values;
- emission standards for mobile sources being too lenient compared to Euro standards;
- incomplete or unreliable emissions inventories, which lead to ineffective actions;
- transboundary and hemispheric air pollution, which cannot be addressed in urban air quality management.

With respect to transboundary and hemispheric air pollution, it is well known that transboundary air pollutants from the Pearl River Delta contribute significantly to the air pollutant levels in Hong Kong (So and Wang, 2003; Lam et al, 2005). The population of Beijing is regularly exposed to high PM concentrations due to sandstorms emerging from desert areas in Western China (Cao et al, 2002). These phenomena need to be addressed if air quality in Asian cities is to be improved.

REFERENCES

APMA/CAI-Asia (2004) *A Strategic Framework for Air Quality Management in Asia*, Air Pollution in the Megacities of Asia (APMA), Stockholm Environment Institute, Clean Air Initiative for Asian Cities, Asian Development Bank, United Nations Environment Programme, World Health Organization, Korea Environment Institute, Seoul, www.cleanairnet.org/caiasia/1412/articles-58180_StrategicFramework.pdf accessed in March 2006

Bell, M. L. and Davis, D. L. (2001) 'Reassessment of the Lethal London Fog of 1952: Novel Indicators of Acute and Chronic Consequences of Acute Exposure to Air', *Pollution Environmental Health Perspectives*, vol 109, (suppl 3), pp389–394

Cao, L., Tian, W., Nu, B., Zhang, Y. and Wang, P. (2002) 'Preliminary Study of Airborne Particulate Matter in a Beijing Sampling Station by Instrumental Neutron Activation Analysis', *Atmospheric Environment*, vol 36, pp1951–1956

Chang, G., Pan, X., Xie, X. and Gao, Y. (2003) 'Time-series Analysis on the Relationship between Air Pollution and Daily Mortality in Beijing' (in Chinese), *Wei Sheng Yan Jiu*, vol 32, pp565–568

Chang, C. C. S. S., Ho, S. C. and Yang, C. Y. (2005) 'Air Pollution and Hospital Admissions for Cardiovascular Disease in Taipei, Taiwan', *Environmental Research*, vol 98, no 1, pp114–119

Chew, F. T., Goh, D. Y., Ooi, B. C., Saharom, R., Hui, J. K. and Lee, B. W. (1999) 'Association of Ambient Air-pollution Levels with Acute Asthma Exacerbation among Children in Singapore', *Allergy*, vol 54, no 4, pp320–329

Clapp, B. W. (1994) *An Environmental History of Britain Since the Industrial Revolution*, Harlow, Longman

Cropper, M. L., Simon N. B., Alberini, A. and Sharma, P. K. (1997) *The Health Effects of Air Pollution in Delhi, India*, PRD Working Paper No. 1860, www.worldbank.org/nipr/work_paper/1860/ accessed in February 2006

Elsom, D. (1996) *Smog Alert – Managing Urban Air Quality*, London, Earthscan

EU (1999) 'Council Directive 1999/30/EC of 22 April 1999 relating to Limit Values for Sulphur Dioxide, Nitrogen Dioxide and Oxides of Nitrogen, Particulate Matter and Lead in Ambient Air', *Official Journal* L 163, 29/06/1999, pp004–0060, http://europa.eu.int/eur-lex/lex/LexUriServ/LexUriServ.do?uri=CELEX:31999L0030:EN:HTML accessed in March 2006

Haq, G. (1997) *Towards Sustainable Transport Planning*, Aldershot, Avebury

Infante-Rivard, C. (1993) 'Childhood Asthma and Indoor Environmental Risk Factors', *American Journal of Epidemiology*, vol 137, pp834–844

Kamat, S. R. (2000) 'Mumbai Studies of Urban Air Pollution and Health Resulting Synergistic Effects of SPM, SO_2 and NO_x', International Conference on Environmental and Occupational Respiratory Diseases, Indian Toxicological Research Centre, Lucknow, India

Kan, H. and Chen, B. (2003a) 'A Case-crossover Study of Ambient Air Pollution and Daily Mortality in Shanghai', *Journal of Occupational Health*, vol 45, pp119–124

Kan, H and Chen, B. (2003b) 'Air Pollution and Daily Mortality in Shanghai: A Time-series Study', *Archives of Environmental Health*, vol 58, no 6, pp360–367

Kan, H., Jia, J., Chen, B. (2004a) 'A Time-series Study on the Association of Stroke Mortality and Air Pollution in Zhabei District, Shanghai', *Wei Sheng Yan Jiu*, vol 33, no 1, pp36–38

Kan, H., Jia, J. and Chen, B. (2004b) 'The Association of Daily Diabetes Mortality and Outdoor Air Pollution in Shanghai, China', *Journal of Environmental Health*, vol 67, no 3, pp21–26

Koe, L. C. C., Arellano, A. F. and McGregor, J. L. (2001) 'Investigating the Haze Transport from 1997 Biomass Burning in Southeast Asia: Its Impact on Singapore', *Atmospheric Environment*, vol 35, pp2723–2734

Lahiri, T. (2004) 'Air Pollution Related Cellular Changes in the Lung of Kolkata and Delhi', paper presented at the conference 'The Leapfrog Factor: Towards Clean Air in Asian Cities', Centre for Science and Environment, New Delhi, 30 April 2004

Lam, K. S., Wang, T. J., Wu, C. L. and Li, Y. S. (2005) 'Study on an Ozone Episode in Hot Season in Hong Kong and Transboundary Air Pollution over Pearl River Delta Region of China', *Atmospheric Environment*, vol 39, pp1967–1977

Lewis, S. A., Corden, J. M., Forster, G. E. and Newlands, M. (2000) 'Combined Effects of Aerobiological Pollutants, Chemical Pollutants and Meteorological Conditions on Asthma Admissions and A&E Attendances in Derbyshire UK, 1993–1996', *Clinical and Experimental Allergy*, vol 30, pp1724–1732

Mashelkar, R. A., Biswas, D. K., Mathur, O. P., Natarajan, R., Nyati, K. P., Singh, D. V. and Singhal, S. (2002) 'Report of the Expert Committee on Auto Fuel Policy', Government of India, New Delhi, http://petroleum.nic.in/ accessed in February 2006

Piver, W. T., Ando, M., Ye, F. and Portier, C. J. (1999) 'Temperature and Air Pollution as Risk Factors for Heat Stroke in Tokyo, July and August 1980–1995', *Environmental Health Perspectives*, vol 107, pp911–916

PRD (2005) '2005 World Population Data Sheet', Population Reference Bureau, Washington DC, www.prb.org/Template.cfm?Section=PRB&template=/Content/ContentGroups/Datasheets/2005_World_Population_Data_Sheet.htm accessed in March 2006

Reuther C. (2000) 'Winds of Change: Reducing Transboundary Air Pollutants', *Environmental Health Perspectives*, vol 108, ppA170–A175, http://ehp.niehs.nih.gov/docs/2000/108-4/focus-abs.html accessed in February 2006

Shankar, P. R. and Rao, G. R. (2002) 'Impact of Air Quality on Human Health: A Case of Mumbai City, India', paper presented at the IUSSP Regional Conference on Southeast Asia's Population in a Changing Asian Context, 10–13 June 2002,

Bangkok, Thailand, www.iussp.org/Bangkok2002/S09Shankar.pdf accessed in February 2006

So, K. L. and Wang, T. (2003) 'On the Local and Regional Influence on Ground-level Ozone Concentrations in Hong Kong', *Environmental Pollution*, vol 123, pp307–317

Tamura, K., Jinsart, W., Yano, E., Karita, K. and Boudoung, D. (2003) 'Particulate Air Pollution and Chronic Respiratory Symptoms among Traffic Policemen in Bangkok', *Archives of Environmental Health*, vol 58, pp201–207

Torres, E., Ronald, D. Subida, R. D., Gapas, J. L., Sarol, J. N., Villarin, J. T., Vinluan, R. J. N., Ramos, B. M. and Quirit, L. L. (2004) *Public Health Monitoring: A Study Under the Metro Manila Air Quality Improvement Sector Development Program (MMAQISDP)*, Manila, Asian Development Bank

UNEP/WHO (1992) *Urban Air Pollution in Megacities of the World*, United Nations Environment Programme/World Health Organization, Oxford, Blackwell

Wang, T., Poon, C. N., Kwok, Y. H. and Li, Y. S. (2003) 'Characterizing the Temporal Variability and Emission Patterns of Pollution Plumes in the Pearl River Delta of China', *Atmosphere Environment*, vol 37, pp3539–3550

Whitelegg, J. and Haq, G. (2003) *An Earthscan Reader in World Transport Policy and Practice*, London, Earthscan

WHO (2000) *Guidelines for Air Quality*, WHO/SDE/OEH/00.02, World Health Organization, Geneva, http://whqlibdoc.who.int/hq/2000/WHO_SDE_OEH_00. 02_pp1–104.pdf, http://whqlibdoc.who.int/hq/2000/WHO_SDE_OEH_00.02pp 105–190.pdf accessed in February 2006

WHO (2005) 'WHO Air Quality Guidelines Global Update 2005', Report of a working group meeting, Bonn, Germany, 18–20 October 2005, World Health Organization, Geneva, www.cepis.ops-oms.org/bvsea/fulltext/guidelines05.pdf accessed in April 2006

WHO/UNEP/MARC (1996) *Air Quality Management and Assessment Capabilities in 20 Major Cities*, UNEP/DEIA/AR.96.2/WHO/EOS.95.7, Nairobi, Kenya, United Nations Environment Programme

Wong, T. W., Lau, T. S., Yu, T. S., Neller, A., Wong, S. L., Tam, W. and Pang, S. W. (1999) 'Air Pollution and Hospital Admissions for Respiratory and Cardiovascular Diseases in Hong Kong', *Occupational and Environmental Medicine*, vol 56, pp679–683

Wong, T. W., Yu, T. S. and Tam, W. (2000) 'Research on Air Pollution and Health in Hong Kong', *Asia Pacific Journal of Public Health*, vol 12, Suppl, S45–547

Wong, C. M., Ma, S., Hedley, A. J. and Lam, T. H. (2001) 'Effect of Air Pollution on Daily Mortality in Hong Kong', *Environmental Health Perspectives*, vol 109, pp335–340

Wong, C. M., Atkinson, R. W., Anderson, H. R., Hedley, A. J., Ma, S., Chau, P. Y. and Lam, T. H. (2002) 'A Tale of Two Cities: Effects of Air Pollution on Hospital Admissions in Hong Kong and London Compared', *Environmental Health Perspectives*, vol 110, pp67–77

Xu, X., Gao, J., Dockery, D. W. and Chen Y. (1994) 'Air Pollution and Daily Mortality in Residential Areas of Beijing, China', *Archives of Environmental Health*, vol 49, no 4, pp216–222

Ye, F., Piver, W. T., Ando, M. and Portier, C. J. (2001) 'Effects of Temperature and Air Pollutants on Cardiovascular and Respiratory Diseases for Males and Females Older than 65 Years of Age in Tokyo, July and August 1980–1995', *Environmental Health Perspectives*, vol 109, pp355–359

five

Trends and Challenges in Managing Urban Air Quality in Asia

INTRODUCTION

In order to determine the current status of AQM in 20 Asian cities, information was collated on current AQM policy and practice in each city and an assessment of AQM capabilities was undertaken using a questionnaire survey. All cities, including those with currently good or excellent management capability, will need to continue to improve their management systems in order to achieve and sustain better air quality.

Compared to the 1990s there has been a general improvement in AQM capabilities and air quality. However, many Asian cities still fail to meet WHO guideline, EU limit values or long-term USEPA standards. SO_2 levels have decreased and have stabilized at relatively low levels in most of the cities examined but are under pressure in those Asian cities which are increasing the use of coal or high-sulphur fuels. Although there has been a slight decrease in TSP and PM_{10} levels, these are still relatively high when compared to EU and USEPA long-term standards. In contrast, NO_2 concentrations have been increasing in highly urbanized areas experiencing rapid motorization, which has also resulted in increasing O_3 concentrations.

The measures taken by the cities to improve their existing AQM capability will determine whether PM_{10} and NO_2 levels can be reduced. Emission control measures can be taken to improve air quality without requiring reliable data, well-developed air quality monitoring networks or emission inventories. These include the adoption and implementation of stringent vehicle emission standards, the introduction of cleaner fuels for mobile and stationary sources and the enforcement of inspection and maintenance programmes. Such measures are particularly beneficial to cities such as Dhaka, Kathmandu and Surabaya with moderate and limited AQM capability.

Despite observable improvements in air quality since 1990, 4500 residents in eight Asian cities (Bangkok, Beijing, Guangzhou, Hong Kong, Kuala Lumpur, Metro Manila, Shanghai and Wuhan) perceive air pollution as worsening (Synovate, 2005). Approximately 88.4 per cent of the respondents surveyed in 2004 felt that air pollution is affecting their daily lives. Respondents from Bangkok felt emotionally affected, 54 per cent felt depressed because of air pollution and 44 per cent of the respondents commented that they were restricting their outdoor activities because of poor air quality. A total of 82 per cent of Metro Manila residents indicated that they are affected by irritation of eyes, nose and throat. These perceptions to some extent reflect the impact of high pollutant concentrations such as PM and NO_2 which have been reported in five cities (Bangkok, Beijing, Hong Kong, Metro Manila and Shanghai). In particular, air quality standards in these countries are above WHO guideline values, EU limit values and USEPA standards.

The perceived failing of measures to manage urban air quality could weaken the willingness of the public and stakeholders to reduce emissions and to comply with air pollution regulations. A negative public perception may undermine the efforts of cities that have excellent AQM capability such as Bangkok, Hong Kong and Shanghai. It could also discourage decision makers from taking action to improve air quality. Equally, public concern over exposure to serious to moderate levels of air pollution could provide an incentive to policy makers to take action to strengthen air management measures.

This chapter presents and discusses a number of trends for the four main components of AQM capability as identified in Chapter 2. It lists common challenges that Asian cities need to address to further improve AQM capability and ultimately achieve better air quality.

AIR QUALITY MONITORING

Due to improvements in air quality monitoring and AQM capability, the status of air quality in Asian cities is becoming increasingly well documented. However, considerable differences remain in the methodology, site selection, frequency and reliability (quality assurance/quality control) of monitoring of air quality and air quality standards. These differences explain why air quality levels and tendencies for one city cannot be compared with those of other cities in Asia nor can they be used as a basis of ranking cities in terms of air quality. Even if these methodological differences did not exist, ranking of cities would still be scientifically unsound. This would be due to differences in air pollutant patterns and the fact that cities not monitoring would elude ranking and cities which monitor air quality would be penalized for the mere fact of monitoring. As a consequence, air quality in each city and its possible improvement can only be analysed by observing tendencies over time.

The number of air quality monitoring stations and the pollutants monitored in Asian cities vary greatly (for example, Dhaka has only one and Colombo only two monitoring stations while Tokyo has 82 stations), which affects spatial representativity. Differences in the number of stations include the:

- availability of funding;
- understanding of the concept 'spatial representativity' and the lack of statistical estimates for the appropriate number of stations under consideration or the residence times of pollutants;
- lack of clarity on how to integrate geographic (land area) demographic (population density) and meteorological considerations (e.g. wind direction, wind speed, topography) in establishing or strengthening their air quality monitoring networks.

While all cities employ a varying number of continuous monitoring stations, a few (e.g. Dhaka, Kathmandu) also use manual monitoring. Continuous monitoring allows for real-time reporting of air quality and, if combined with the presence of a data server that broadcasts real-time air quality information, offers more complete information more quickly in support of AQM. Beijing uses a large number of diffusive (passive) samplers in addition to automatic continuously monitoring devices. Diffusive samplers are easy to handle, cheap and sufficiently accurate for longer exposure times. They allow the monitoring network to achieve good spatial representativity.

In cities such as Dhaka, Hanoi, Kolkata, Mumbai and New Delhi, several agencies are monitoring air pollution. Data are not necessarily harmonized (e.g. Hanoi, Mumbai) and consequently, monitoring results from sites in close vicinity are not always comparable with each other.

There are still considerable differences in the number and type of pollutants measured in the various cities. Since the late 1990s PM_{10} has been monitored routinely in Asian cities while only half of the cities examined appear to have monitored O_3. The decision to monitor additional pollutants is mainly driven by whether or not standards exist for specific pollutants. Following the example of European and US cities, and in view of the recent evidence of health impacts of fine PM and O_3 (WHO, 2003; 2004), Asian cities are expected to intensify efforts to monitor O_3 and $PM_{2.5}$ in the future. However, the level and frequency of monitoring will vary due to differences in priority and budgets allocated for this purpose.

Cities such as Bangkok, Hong Kong and Ho Chi Minh City have begun to differentiate between air quality monitoring at urban sites away from stationary emission sources and roadside monitoring close to mobile sources. The reason for this is that in most Asian countries many people are exposed to air pollution from mobile sources in the immediate vicinity of the road (e.g. street vendors and their customers, policemen, people living along streets in slum areas). This is an approach which should be considered by more cities given the large contribution of mobile sources to air pollution. Roadside air quality monitoring should be part of an integrated AQM approach which has standardized methodologies for air quality monitoring at different urban locations. Such an approach will consider the influence of industrial and transboundary sources in addition to mobile sources and ensure that roadside monitoring results can be viewed in the context of the overall air pollution situation.

A common weakness in air quality monitoring in Asian cities is the quality assurance and control procedures. Rigorous quality assurance plans are usually not developed or implemented. Without robust quality assurance procedures, air quality data are subject to discussion because they may be of unknown quality. Such data cannot provide the basis for a detailed assessment of tendencies in air quality, source apportionment, compliance with standards, and health and environmental impacts. As a consequence policy making and enforcement of emission regulations may be weakened.

Although air quality monitoring has improved in major Asian cities, a number of these cities still do not have the capability to monitor urban air quality adequately. This is especially true for the secondary cities (i.e. cities of fewer than 500,000 people) of which there are estimated to be more than 2000 in Asia.

The cost of air quality monitoring is still an important factor and it cannot be expected that these cities will be able to install continuous air quality monitoring stations. Diffusive monitoring is another technique that can be used as a tool to identify air pollution hotspots. Diffusive samplers also provide a low-cost alternative in the absence of more sophisticated monitoring equipment, provided that the capability exists to undertake reliable measurement and analysis of samples. This makes diffusive monitoring an attractive option for secondary cities in Asia. At present, the diffusive monitoring technique is rarely used for regular monitoring activities in Asian cities. The use of these devices, particularly in cities with moderate or limited capability, could enhance their capability in air quality monitoring.

In the last few years progress has been made in the use of satellite observations for air quality monitoring. Satellite data can aid in the detection, tracking and understanding of pollutant transport by providing observations over large spatial domains and at varying altitudes. Several satellite sensors can detect some types of aerosols, including smoke plumes from fires (Engel-Cox et al, 2003). They can thus provide a basis for deriving emission estimates of events such as forest fires (Neil et al, 2003). Soil moisture can be detected from space and is an essential piece of information for estimating how much dust (a type of PM) is contributing to atmospheric haze. Several satellite missions designed to detect stratospheric O_3 can also provide information on tropospheric O_3 levels (Fishman et al, 2003).

The general use of satellite-based monitoring techniques would allow cost-effective air quality data for a large number of Asian cities. Initially, such data will probably be used for research purposes to determine whether concentration at receptor sites can be adequately estimated. However, it will be some time before such techniques could be used for regulatory purposes.

The Asian Regional Research Program on Environmental Technology (ARRPET) and the International Atomic Energy Agency programme on source apportionment of PM using nuclear analytical techniques have been monitoring air pollution for research purposes (Oanh et al, 2004; IAEA, 2004). The results of these initiatives, which often involve passive sampling, until recently have not been used to validate the results of routine ambient air quality monitoring. The results of such research programmes are rarely communicated to local governments and the public.

Assessing the impacts of urban air pollution

The first step in the implementation of any AQM strategy involves measuring the concentration of air pollutants and determining the current level of air quality using air quality standards. Policy makers will require additional information on the impacts of air pollution to support policy making. Until recently, literature on health impacts of air pollution in Asia has not being as abundant as in developed counties and this is still the case in the least developed countries in Asia (WHO, 2000). However, progress has been made in determining the health impacts of air pollution in Asia. The Public Health and Air Pollution in Asia (PAPA) programme of the Clean Air Initiative for Asian Cities (CAI-Asia) consolidated for the first time existing health impact studies. PAPA has initiated a series of coordinated health impacts studies in three Asian countries (HEI, 2004). Research is currently being undertaken to examine the links between air pollution, poverty and health impacts in Ho Chi Minh City (ADB, 2005).

The evidence basis for other environmental impacts such as reduced crop yields and damage to buildings and infrastructure is still very limited. There is a need to expand air pollutant monitoring networks to cover agricultural and forest areas. Monitoring data of ground-level SO_2, NO_2 and O_3 are, however, scarce in most rural areas. Concentrations of O_3 are approaching phytotoxic levels in some parts of Asia (Ashmore et al, 2005). There is a severe shortage of reliable local exposure–yield data and the influence of the mixture of pollutants on exposed vegetation. There is a need to systematize past and ongoing research efforts and to broaden the number of organizations involved in such research (Ashmore et al, 2005).

Chapter 1 discussed the economic impact of air pollution in selected Asian cities. However, considerable gaps still exist in determining the economic cost of urban air pollution in Asia. Methodological weaknesses make it difficult to compare economic impact studies between countries and cities. Notwithstanding the gaps in coverage and methodological weaknesses, it is clear that the costs of air pollution to society are high and outweigh the costs of air pollution abatement. An EC study, which estimated health costs in 2000 and 2020, identified benefits between 89 and 193 billion euro if current EU air pollution legislation is implemented in the 25 EU member states in 2020 (AEAT, 2005). The US Office of Management and Budget estimated the benefits of enforcing four AQM-related policies to be US$101–119 billion per year from 1992 to 2002 compared to the cost of US$8–8.8 billion per year (USOMB, 2003). The policies included two rules limiting PM and NO_x emissions from heavy-duty highway engines, the Tier 2 rule limiting emissions from light-duty vehicles, and SO_2 limits as part of the acid rain provisions in the 1990 amendments to their Clean Air Act.

Increased political and public awareness of the health and environmental impacts of air pollution and their costs in relation to the costs of air pollution abatement will be instrumental in advancing urban AQM in Asia.

Data assessment and availability

Air quality reporting is as important as monitoring itself, as it provides policy makers and stakeholders responsible for AQM with the information on which to base their efforts to reduce air pollution. This includes making available average hourly, daily, weekly, monthly and yearly averages of air quality data and their confidence intervals. In most of the Asian cities examined here information on air quality levels is now made available as an aggregate form via a daily Air Pollution Index (API) (CAI-Asia, 2006a). Following the USEPA Pollutant Standard Index (PSI), APIs have been developed as an instrument to communicate air quality levels in a more easily understandable manner to the general public and alert them when pollutant concentrations reach levels harmful to human health.

Policy makers and researchers cannot develop integrated AQM strategies based solely on APIs and averaged datasets. Policy makers need information on compliance with standards, potential impacts of air pollution, and the costs and benefits of control actions. Researchers need to have access to individual reliable monitoring data in order to be able to conduct research on health and environmental impacts, and to estimate the economic cost of air pollution. Similarly, air quality modellers need information on air quality in order to validate their models; researchers conducting source apportionment studies need to have access to detailed and reliable air quality data. The reality in many of the Asian cities examined here, as well as in other Asian cities, is that individual data are often not readily available and, if they are, it is at a financial cost and often with a considerable time delay.

There is a need to improve data interpretation. This includes the understanding of compliance testing, the discretional use of meteorological data and the careful application of statistical methods. In short, to interpret data in a suitable context in order to provide decision makers with useful information and knowledge.

EMISSION INVENTORIES

Source apportionment is the estimation of the contribution of different sources to an observed air pollutant situation. It is applied when air quality standards are exceeded and there is a need to understand the contribution of individual sources. Source apportionment can provide support to modellers in the development of control options and to policy in the formulation of clean air implementation plans. It allows the cost-efficiency of control options to be estimated. If air quality monitoring is imperfect or invalid, as is the case in a large part of Asian cities, this will limit the validity of source apportionment.

Emission inventories estimate emissions directly at the source using stack monitoring, input–output analysis or emission factors. They enable policy makers and regulators to prioritize and target the control of emissions from major sources of emissions, provided emissions are reliably estimated. If emissions inventories are imperfect, fragmented or non-existent, this will limit the potential of applying the full range of integrated AQM measures.

In most of the Asian cities examined, reliable emissions inventories are either lacking or contradictory or cover only part of all sources of air pollutants, for example by assuming that mobile emissions are the major source and industrial and transboundary sources play only a subordinate role. Mostly they are ad hoc estimates conducted by international or national organizations on a one-time level or within projects independent of a rational AQM system. Emissions inventories often do not undergo procedures of quality assurance and quality control. This limits their use in policy making or in the evaluation of the implementation of policy measures.

In Asian cities, emission estimates are compiled for mobile sources or for power plants rather than providing a comprehensive assessment of all sources of air pollution, which would deserve the name of an emission inventory. This adds to the difficulties many cities and countries in Asia face in accurately quantifying the contribution made by different sources to urban air pollution. The contribution to air pollution of area sources such as road dust and open burning, inside and outside the cities, is of particular concern, especially with respect to fine PM. Additional studies and the application of better methodologies are required to be able to arrive at information on which sound policy making can be based.

Asian cities are becoming increasingly aware of the influence of emissions which originate from outside the city. Emissions from the Pearl River Delta have resulted in Hong Kong residents being exposed to increased air pollution levels (Wang et al, 2003) while the sandstorms from desert areas in China have contributed to smog in Beijing and other northeastern Chinese cities (Sun et al, 2004; Zhang et al, 2004; NASA, 2004; *China Daily*, 2006). In addition, regional sources of pollution such as the Southeast Asian haze (Wheeler, 1998) and the atmospheric brown cloud (UNEP, 2000a) play a significant role in urban air pollution. These regional processes have contributed to urban pollution levels, which are already exceeding air quality standards.

There is also a growing concern over the impact of emissions which originate from Asia on air quality in other parts of the northern hemisphere (e.g. Jaffe et al, 2003; Iino et al, 2004; Vingarzan, 2004). This rising awareness of the importance of regional, transboundary and hemispheric transport of air pollutants has resulted in an increase in scientific studies which will help to develop a better understanding of urban air in cities exposed to long-distance air pollution (Lee et al, 2004; Kar and Takeuchi, 2004; Grousset and Biscaye, 2005).

Within Europe and other parts of the developed world more attention is now being given to the contribution made by shipping (EC, 2005) and aviation (ATAG, 1999; CGER, 1999) to air pollution. Methodologies are being developed to study these sources (e.g. Advanced Emission Model for Aviation). The results of studies on aviation and shipping are of considerable relevance to Asia where the aviation sector is rapidly growing and shipping is making a significant contribution to the economic development of coastal cities (Jelinek et al, 2004).

Emissions inventories in Asian cities often use emissions factors that are not specific to the technology commonly used in the country or the city for which the emission inventory has been compiled. Often default emission factors are used, which were developed in Europe or the US. Similarly, activity data, the second main element of emission inventories, are often weak in Asian cities due to the poor capacity in government agencies responsible for recording vehicle numbers or industrial activity.

AIR QUALITY MANAGEMENT STRATEGIES

Since the early 1980s a number of measures have been taken to address air pollution problems in Asia. Measures to address urban air pollution intensified in the mid-1990s when the World Resources Institute identified several Asian cities among the world's worst polluted cities (Davis and Saldiva, 1999). These measures have been helpful in stabilizing pollution levels and in some cases reversing air quality trends (WHO, 2001). However, as discussed, considerable challenges still exist that have to be overcome before many Asian cities can enjoy better air quality.

Most Asian countries have adopted ambient air quality standards. However, such standards in Asia are not harmonized and vary considerably. Countries tend to follow WHO guidelines in setting the standards and in the case of PM_{10}, USEPA standards. Where standards deviate from WHO guideline values, EU limit values and USEPA standards, Asian ambient standards tend to be more lenient. This is particularly the case for the 24-hour standards for PM_{10}, which are greater than the EU limit by a factor of two or more in Bangladesh, China, Hong Kong, Indonesia, Nepal, Philippines, Republic of Korea, Singapore, Taipei and Thailand. It also applies to the 1-year NO_2 standards in Bangladesh, Indonesia, Republic of Korea, Taipei, China and Vietnam. There is also considerable variation in averaging times under the ambient standards established by Asian countries compared to the WHO guideline values, EU limit values and USEPA guidelines.

In most Asian countries, PM_{10} is the main pollutant of concern and currently concentrations are well above EU annual and 24-hour limits and the USEPA annual standard. The WHO determined that no threshold limit could be established for PM_{10} although a guideline value of 20 $\mu g/m^3$ has recently been set (WHO, 2005). The EU has set a 24-hour limit much lower than that of the USEPA. As discussion on the need to tighten the PM_{10} standards in Asia intensifies, regulators will face a dilemma whether to set strict standards for fine PM as low as that of the EU or to adopt more lenient standards, which can be achieved with current AQM capability.

A certain amount of confusion exists among regulators on whether to set air quality standards for NO_x (the sum of NO and NO_2) or NO_2. However, NO is known to have little effect on human health or the environment (WHO, 2000). Similarly, regulators in Asian countries have adopted different approaches in setting averaging times for measuring O_3 concentrations. The harmonization of approaches between Asian countries in standard setting can assist in promoting the inter-comparability of air quality monitoring results for individual pollutants across cities and countries.

Although ambient standards have been established in most Asian countries, they are not necessarily used in all cities and implemented as part of an integrated AQM system comprising air quality monitoring, source identification, emission inventories, emission control strategies and health impact assessment. In Japan, Sri Lanka and Vietnam, long-term standards for PM_{10}, NO_2 and SO_2 have not been adopted or enforced, which limits the estimation of long-term health impacts. The same applies to Nepal (no long-term PM_{10} standard), Philippines and Thailand (no long-term NO_2 standard), and Republic of Korea (no long-term NO_2 or SO_2 standards). For example, in Bangladesh, China and the Philippines and, until recently, Vietnam, O_3 standards exist but routine air quality monitoring activities do not include O_3. It is not certain whether the O_3 levels meet the set standards. In contrast, Indian cities monitor O_3 in spite of the fact that India has not adopted an O_3 standard.

The majority of Asian countries have adopted emission standards for motor vehicles and stationary sources of pollution. Motor vehicle emission standards, which in most cases are linked to fuel quality standards, are in nearly all cases based on the European emission standards. In recent years, the legislation to tighten emission standards for new vehicles taken by several Asian countries has reduced the time lag between Europe and Asia from approximately 8–10 years in 1995 to 4–6 years in 2005. A critical factor in the implementation of more stringent vehicle emission standards in Asia will be the success of Asian countries in introducing cleaner fuels, especially reducing the sulphur content of fuel. Although several Asian countries have developed, or are in the process of developing cleaner fuels, questions remain on the speed at which refinery modifications can be undertaken to produce ultra-low sulphur fuels. A potential common aim could be the adoption of Euro IV emission standards by 2010 (ADB, 2003; JAMA, 2006).

In contrast to vehicle emission standards, less transparency exists on emission standards for stationary sources and the manner in which such standards will develop over the next 5–10 years in Asia. An area of particular concern is the increase in coal consumption for power generation and other industrial purposes. This has resulted in considerable research and development activities into cleaner coal technology (UIC, 2004; USDOE, 2002). The emergence of such technology might lead to the setting of stricter emission standards for stationary sources to ensure faster adoption.

Integrated approach to AQM

In contrast to the past, when major causes of air pollution occurred sequentially rather than simultaneously, Asian cities are now faced with the pressure of a combination of driving forces (e.g. motorization, industrialization and energy demand). To deal with such challenges and to efficiently manage air quality in the cities, the four AQM indices (measurement; data assessment and availability; emissions estimates; and management capacities) as well as other related environmental concerns of the cities should be integrated into a strategic framework for AQM. An integrated approach to AQM is comprehensive and aims to facilitate the setting of air quality priorities and provide direction on supportive institutional development and capability enhancement. Such an approach has been formulated in the Strategic Framework for AQM in Asia (APMA/CAI-Asia, 2004). The Strategic Framework provides a broad high-level approach targeted at government, industry, media, academia, non-governmental organizations and the general public. It outlines the key challenges existing in AQM in Asia and provides recommendations with respect to different components of a comprehensive AQM system. By taking a strategic approach to AQM some of the challenges highlighted can be addressed to allow effective AQM for better air quality in Asia. This strategic and integrated AQM approach would incorporate the following:

- Developing clean air implementation plans as formulated in the Strategic Framework. This includes addressing the greatest challenges in Asian cities such as emission inventories and health and environmental impact assessments.
- Extending dedicated air quality legislation to other sectors, involving all stakeholders, and strengthening the use of economic instruments.
- Emphasizing transport demand management and transport planning. This would include reducing the need to travel for essential services, providing attractive and efficient public transport systems, developing an urban environment that is clean, safe and more conducive to walking and cycling.
- Increasing the use of land use planning tools in AQM (e.g. zoning) and taking into consideration air quality information in roads and industrial site planning.
- Tightening air quality standards towards the EU limit values and/or WHO guideline values.
- Increasing political and public awareness of the health and environmental impacts of air pollution and their costs in relation to the costs of air pollution abatement.
- Promoting more public health and environmental studies in the framework of environmental impact assessment.

Urban development in Asian cities has in most cases been guided by economic development considerations. Concerns about environmental deterioration and air pollution usually had to take second place. This has resulted in a management approach with a heavy emphasis on technological solutions and limited attention for preventive approaches. It is interesting to note that the cities in the study with the highest scores in terms of AQM capability – Hong Kong, Singapore, Taipei and Tokyo – are also cities with the most well-developed public transport systems and land use planning policies.

For Asian cities to be able to meet air quality goals they will have to integrate urban planning, land use planning, transport planning and air quality management. Such approaches are already becoming increasingly common in some parts of Western Europe (EEA, 2005). In Asian cities there is a growing interest in considering new approaches to revitalize cities by providing sustainable mobility such as the bus rapid transit (BRT) Systems which are currently operating in Beijing, Jakarta, Kunming, Taipei, Seoul and several Japanese cities. BRT systems are under construction or planned in Ahmedabad, Bangalore, Bangkok, Hanoi, Metro Manila, New Delhi, Pune and several Chinese cities including Shanghai, Shenyang and Wuhan (BigNews, 2004; *The Nation*, 2005; CAI-Asia, 2005). Better transport demand management and transport planning will help to develop an urban environment which is cleaner, safer and more conducive to walking and cycling.

To strengthen a preventive and integrated approach to AQM, the use of economic instruments will have to be widely applied in managing stationary and mobile air pollution. Singapore led the way by being one of the first countries to introduce congestion charges, which has been followed by Trondheim, Norway and London, UK. Congestion charging has proved to be an effective measure to reduce private vehicle use, improve public transport and reduce emissions and congestion (BBC News, 2004; BBC, 2006).

There is now increasing experience of emission trading for SO_2, NO_x and GHG to control emissions from stationary sources (Ellerman et al, 1997; 2003; Swift, 2000; NRTEE, 2004). A further improvement of AQM capability in air quality monitoring, emission inventories and source apportionment studies in Asian cities will make it easier to adopt wide-scale emission trading in Asia.

Urban air pollution and global climate change have similar drivers: urbanization, motorization and increased energy use. There is also growing evidence of the ancillary benefits of GHG reduction for air pollution (Alcamo et al, 2002; van Vuuren et al, 2006). This makes it relevant for decision makers not only to consider air quality in setting policies and regulations on urban air quality but at the same time to consider what benefits for air pollution can be achieved with policy measures to mitigate global climate change. This will help to mobilize support and resources for the adoption and implementation of such policies.

A frequent complaint about AQM in Asian cities is the lack of enforcement of air quality regulations and emission standards (APMA/CAI-Asia, 2004). Information on the lack of enforcement is mostly anecdotal and few systematic studies have been undertaken. However, it is clear that considerable scope for improvement exists in enforcing standards. This has a direct result in terms of reducing emissions. It is also reasonable to expect that stronger enforcement of regulations will have a more effective impact in preventing emissions. Studies from Jakarta indicate that the wide-scale adoption of preventive inspection and maintenance programmes for in-use vehicles can help to reduce emissions by 50 per cent. Emissions of HC have been reduced by 27 per cent for gasoline vehicles and emissions of soot by 56 per cent for diesel vehicles. An improved fuel efficiency of 3–4 per cent was also observed as a result of the programme (Pramono and Heuberger, 2001).

In some cases the poor enforcement of air quality standards and regulations has resulted in citizens launching court cases against the government. Some of these cases have been instrumental in forcing governments to take action, which has resulted in considerable improvements in air quality. A well-known example is that of the 1998 Indian Supreme Court direction to the city of New Delhi. This required the city to undertake a number of measures which included improving public transport; banning the use of buses older than eight years; converting the entire diesel bus fleet to CNG; and replacing all pre-1990 autos and taxis with new ones using cleaner fuel by March 2000. The court decision was a consequence of the 'Right to Clean Air' campaign of the Centre of Science and Environment intended to improve air quality (CSE, 2006). Although the implementation of the court decision was delayed by industry and government, it finally resulted in less polluted air in New Delhi. Legal action initiated by citizens and non-governmental organizations is therefore a potential instrument to force several Asian cities to take action, especially in those cities which have limited AQM capability.

AQM is still a relatively new concept in most of the cities examined and a weak institutional capability contributes to the relatively low AQM capability in some cities. Asian countries and cities will have to increase the numbers of staff working on all aspects of AQM. They will have to develop a programme of continuous training to ensure that staff are familiar with new technologies and methodologies. However, adding more staff and better training and the provision of equipment will not be sufficient to generate the required institutional capability to manage air quality successfully. In many cases there will be a need to rearrange institutional mandates or create new mandates to ensure better coordination between national and local governments as well as coordination between different sectors in government and between government, private sector and civil society.

DEVELOPING A REGIONAL APPROACH TO IMPROVE URBAN AIR QUALITY IN ASIA

Although each city's air pollution problems are to some extent unique and require specific AQM strategies, it is becoming increasingly apparent that Asian cities are experiencing similar problems, for example stagnant or increasing PM, NO_x and O_3 concentrations. Urban air pollution problems are being further exacerbated by emissions from sources outside the city, for example transboundary air pollution such as dust and sandstorms from mainland China or haze from fires in South East Asia.

A regional approach to air pollution allows an understanding of the causes and impacts of air pollution and maximizes resources including capacity building and funding opportunities. Asian countries are increasingly working together to address air pollution problems. Programmes such as the Acid Deposition Monitoring Network in East Asia (EANET, 1999) and the Atmospheric Brown Cloud Project (UNEP, 2000a), and declarations and agreements such as the Malé Declaration on Control and Prevention of Air Pollution and its Likely Transboundary Effects for South Asia (UNEP, 2000b) and the ASEAN Agreement on Transboundary Haze (ASEAN, 2002) play a critical role in developing a better scientific understanding of transboundary air pollution in the Asia. The UNECE Task Force on Hemispheric Transport of Air Pollution initiated a programme to examine the hemispheric transport of air pollutants (UNECE, 2004) as discussed in Chapter 1. CAI-Asia activities aimed at the exchange of information, knowledge and capacity building have raised the profile of urban air pollution in the region.

These programmes and activities demonstrate the viability of regional approaches to improving AQM in Asian countries. They all contribute to improving scientific understanding, which is the first step in formulating common regional management strategies for the mitigation of air pollution impacts. Provided a conducive political context exists, there is no reason why Asian governments cannot begin to replicate the successful intergovernmental cooperation, which resulted in reduced transboundary air pollution in Europe. Although Europe still has some way to go before it recovers from the legacy of past decades of acid deposition, acidification is expected to continue declining due to the integrated approach taken to address acidification, eutrophication and ground-level ozone (EEA, 2005). Pending the establishment of such intergovernmental cooperation in Asia, regional initiatives such as those introduced by UNEP, CAI-Asia and others can assist in influencing policies at the national and city level and to ensure that there is investment in support of AQM goals set by governments (CAI-Asia, 2006b).

CONCLUDING REMARKS

Despite considerable progress being made to clean the air in Asian cities, air pollution continues to pose a threat to human health, environment and the quality of life in cities. The concoction of increasing migration, motorization and uncontrolled urban growth has all contributed to the intensification of air pollution, which currently poses a significant challenge to Asian cities.

This assessment of the current status, challenges and management in urban air pollution in 20 Asian cities provides a benchmark against which future initiatives and progress in AQM can be assessed. The identification of the stage of development in terms of AQM capability can assist cities in setting priorities and developing strategies to strengthen their AQM capability. Cities with a relatively low AQM capability need to focus on establishing or strengthening continuous and diffusive air quality monitoring systems and implementing basic control strategies. Cities with higher AQM capability should focus on improving emission data. In the development of air pollution control strategies, they should aim to address demand management in transport and integrate local air pollution measures with transboundary air pollution and GHG abatement. All cities will need to ensure that their AQM systems not only manage the traditional criteria pollutants such as CO, NO_x, SO_2, O_3, TSP and PM_{10}

Table 5.1 *Challenges in AQM of Asian cities*

Air quality policies	Several bodies or ministries are sometimes responsible for maintaining acceptable air quality hampering, sectoral integration if coordination is limited; Absence of appropriate review mechanisms; Regional differences in regulation of emission sources; Deficiencies in setting air quality standards: Lack of: ■ stakeholder participation ■ up-to-date emission standards ■ monitoring and quantitative data on air quality and its impact on human health and the environment ■ up-to-date air quality standards.
Air quality governance	Prevalence of ad hoc awareness raising with a focus on raising alarm; Deficiencies in information dissemination; Design and implementation of AQM strategies often based on incomplete knowledge; Potential for misinterpretation of air quality reporting and information.
Emissions	Lack of: ■ emission inventories and quality assured emission data ■ source apportionment ■ periodical update of emission standards ■ regional harmonization of emission standards.
Air quality modelling	Lack of: ■ quality assured emission data ■ source apportionment experience ■ regional harmonization of dispersion models.
Air quality monitoring	Absence of: ■ and/or limited coverage of outdoor air quality monitoring systems ■ periodic review of air quality monitoring issues ■ or limited existence of baseline data ■ or poor quality data and limited dissemination ■ focus on control and quality assurance of monitoring programmes. Lack of: ■ standard operating procedures for monitoring, for data analysis and presentation ■ harmonization of monitoring networks and devices ■ monitoring of transboundary air pollution ■ hotspot monitoring.
Health, environmental and economic risk assessments	Lack of long-term studies of health, environmental and economic impacts due to air pollution; Poor information and assessment of health, environmental and economic impacts of air quality; Insufficient representativity of monitoring sites for actual exposure of humans and the environment.

Source: APMA/CAI-Asia (2004)

but also fine PM, which is monitored as $PM_{2.5}$. In addition, all countries should review their air quality standards in view of the EU limit values and the new WHO guideline values (EC, 1999; 2000; 2002; WHO, 2000; 2005). Table 5.1 summarizes the main challenges faced by Asian cities.

Although other global studies of AQM have included Asian cities (UNEP/WHO, 1992; WHO/UNEP/MARC, 1996), this is the first of its kind to provide a systematic and comparative analysis of 20 Asian cities. Periodic assessments of AQM allow historical patterns to be distinguished. By taking a historical perspective the relatively strong increase in AQM capability in China – and also other countries – becomes apparent. By monitoring the progress made in developing AQM capabilities, learning from past experience and identifying future priorities, national and local authorities can be assisted in achieving a more sustainable urban environment and a better quality of life for the millions of people living in Asian cities.

REFERENCES

ADB (2003) 'Reducing Vehicle Emission in Asia. Policy Guidelines for Reducing Vehicle Emissions in Asia', Asian Development Bank, Manila, www.adb.org/documents/guidelines/Vehicle_Emissions/reducing_vehicle_emissions.pdf accessed in April 2006

ADB (2005) 'Air Pollution, Poverty and Health Effects in Ho Chi Minh City', Asian Development Bank Technical Assistance VIE- 4714, Manila www.adb.org/Documents/ADBBO/AOTA/39076012.ASP accessed in March 2006

AEAT (2005) 'CAFÉ CBA: Baseline Analysis 2000 to 2020', AEA Technology: Service contract for carrying out cost benefit analysis of air quality related issues, in particular in the Clean Air for Europe (CAFÉ) Programme. AEA Technology Environment, Didcot, UK. www.muenchen.de/vip8/prod1/mde/_de/rubriken/Rathaus/70_rgu/04_vorsorge_schutz/luft/luftqualitaet/feinstaub/pdf/CAFE_CBA_baseline_analysis_januar2005.pdf accessed in April 2006

Alcamo, J., Mayerhofer, P., Guardans, R., van Harmelen, T., van Minnen, J., Onigkeit, J., Posch, M. and de Vries, B. (2002) 'An Integrated Assessment of Regional Air Pollution and Climate Change in Europe: Findings of the AIR-CLIM Project', *Environmental Science and Policy*, vol 5, pp 257–272

APMA/CAI-Asia (2004) 'A Strategic Framework for Air Quality Management in Asia', Air Pollution in the Megacities of Asia (APMA), Stockholm Environment Institute, Clean Air Initiative for Asian Cities, Asian Development Bank, United Nations Environment Programme, World Health Organization, Korea Environment Institute, Seoul, www.cleanairnet.org/caiasia/1412/articles-58180_StrategicFramework.pdf accessed in March 2006

ASEAN (2002) 'ASEAN Agreement on Transboundary Haze Pollution', Association of Southeast Asian Nations, Jakarta, www.aseansec.org/pdf/agr_haze.pdf also see: http://en.wikipedia.org/wiki/ASEAN_Agreement_on_Transboundary_Haze_ Pollution accessed in April 2006

Ashmore, M., Emberson, L. and Murray, F. (2005) *Air Pollution Impacts on Crops and Forests: A Global Assessment*, London, Imperial College Press

ATAG (1999) 'Aviation and the Environment', Air Transport Action Group, Geneva, www.atag.org/files/AvEv-150311A.pdf accessed in April 2006

BBC (2006) 'Case Studies of Other Charging Zones', British Broadcasting Corporation, London, www.bbc.co.uk/london/congestion/cities.shtml accessed in April 2006

BBC News (2004) 'Congestion Charging "A Success"' BBC News, 17 February 2004, British Broadcasting Corporation, London, http://news.bbc.co.uk/1/hi/england/ london/3494015.stm accessed in April 2006

BigNews (2004) 'Bus Rapid Transport Systems Making Inroads in Asia', BigNews Network, 8 December 2004, http://feeds.bignewsnetwork.com/?sid=a662762f4b6 15b64 accessed in April 2006

CAI-Asia (2005) 'Bus Rapid Transit Overview', Clean Air Initiative for Asian Cities, Manila, www.cleanairnet.org/caiasia/1412/article-59592.html accessed in April 2006

CAI-Asia (2006a) 'Ambient Air Quality Monitoring Data in Asia', Clean Air Initiative for Asian Cities, Manila, www.cleanairnet.org/caiasia/1412/article-59646.html accessed in April 2006

CAI-Asia (2006b) 'Clean Air Initiative for Asian Cities' Manila, www.cleanairnet. org/caiasia/1412/channel.html accessed in April 2006

CGER (1999) 'A Review of NASA's Atmospheric Effects of Stratospheric Aircraft Project', Panel on Atmospheric Effects of Aviation, Board on Atmospheric Sciences and Climate, Commission on Geosciences, Environment, and Resources, National Research Council. National Academy Press Washington DC, http://darwin.nap. edu/books/0309065895/html/ accessed in April 2006

China Daily (2006) 'Sanstorm Sweeps NW China Region', 12 March 2006, www. chinadaily.com.cn/english/doc/2006-03/12/content_533283.htm accessed in April 2006

CSE (2006) *The Leapfrog Factor: Cleaning Air in Asian Cities*, New Delhi, India, Centre for Science and Environment

Davis, D. L. and Saldiva, P. H. N. (1999) 'Urban Air Pollution Risks to Children: A Global Environmental Health Indicator', *Environmental Health Notes*, New York, World Resources Institute

EANET (1999) 'Acid Deposition Monitoring Network in East Asia', www.eanet.cc/ accessed in April 2006

EC (1999) 'Council Directive 1999/30/EC of 22 April 1999 relating to Limit Values for Sulphur Dioxide, Nitrogen Dioxide and Oxides of Nitrogen, Particulate Matter and Lead in Ambient Air', *Official Journal*, L 163, 29/06/1999, pp0041–0060, http://europa.eu.int/eur-lex/lex/LexUriServ/LexUriServ.do?uri=CELEX:31999L0030:EN:HTML accessed in April 2006

EC (2000) 'Directive 2000/69/EC of the European Parliament and of the Council of 16 November 2000 relating to Limit Values for Benzene and Carbon Monoxide in Ambient Air', *Official Journal*, L 313, 13/12/2000 pp0012– 0021, http://europa.eu.int/eur-lex/lex/LexUriServ/LexUriServ.do?uri=CELEX:32000L0069:EN:HTML accessed in April 2006

EC (2002) 'Directive 2002/3/EC of the European Parliament and of the Council of 12 February 2002 relating to Ozone in Ambient Air', *Official Journal*, L67, 09/3/2002, pp14–27, http://europa.eu.int/eur-lex/pri/en/oj/dat/2002/l_067/l_0672 0020309en00140030.pdf accessed in April 2006

EC (2005) 'Transport and Environment', European Commission, Brussels, http://europa.eu.int/comm/environment/air/transport.htm accessed in April 2006

EEA (2005) 'The European Environment. State and Outlook 2005', European Environmental Agency, Copenhagen, http://reports.eea.eu.int/state_of_environ ment_report_2005_1/en/tab_content_RLR accessed in April 2006

Ellerman, A. D., Schmalensee, R., Joskow, P. L., Montero, J.-P. and Bailey, E. M. (1997) 'Emissions Trading under the US Acid Rain Program: Evaluation of Compliance Costs and Allowance Market Performance', MIT Center for Energy and Environmental Policy Research, Cambridge, MA, Massachusetts Institute of Technology,

Ellerman, A. D., Joskow, P. L. and Harrison Jr., D. (2003) 'Emissions Trading in the US: Experience, Lessons and Considerations for Greenhouse Gases', prepared for the Pew Center on Global Climate Change, May 2003, www.pewclimate.org/global-warming-in-depth/all_reports/emissions_trading/index.cfm accessed in April 2006

Elsom, D. (1996) *Smog Alert – Managing Urban Air Quality*, London, Earthscan

Engel-Cox, J., Haymet, A. and Hoff, R. (2003) 'Review and Recommendations for the Integration of Satellite and Ground-based Data for Urban Air Quality', Air & Waste Management Association Annual Conference and Exhibition, San Diego, CA

Fishman, J., Wozniak, A. E. and Creilson, J. K. (2003) 'Global Distribution of Tropospheric Ozone from Satellite Measurements Using the Empirically Corrected Tropospheric Ozone Residual Technique: Identification of the Regional Aspects of Air Pollution', *Atmospheric Chemistry and Physics*, vol 3, pp893–907

Grousset, F. E. and Biscaye, P. E. (2005) 'Tracing Dust Sources and Transport Patterns Using Sr, Nd and Pb isotopes', *Chemical Geology*, vol 222, pp149–167

HEI (2004) *Health Effects of Outdoor Air Pollution in Developing Countries of Asia: A Literature Review*, HEI Special Report 15, Boston, MA, Health Effects Institute,

IAEA (2004) 'Air Pollution Summary Report. Thematic Planning Meeting on Monitoring Air Pollution', 7–11 June 2004, Vienna, Austria, www-tc.iaea.org/tcweb/abouttc/strategy/thematic/pdf/summary/Summary_Report_AirPollution.pdf accessed in April 2006

Iino, N., Kinoshita, K., Tupper, A. C. and Yano, T. (2004) 'Detection of Asian Dust Aerosols using Meteorological Satellite Data and Suspended Particulate Matter Concentrations', *Atmospheric Environment*, vol 38, pp6999–7008

Jaffe, D., McKendry, I., Anderson, T. and Price, H. (2003) 'Six "New" Episodes of Trans-Pacific Transport of Air Pollutants', *Atmospheric Environment*, vol 37, pp391–404

JAMA (2006) 'Japan Automobile Manufacturers Association Newsletter', vol 18, February 2006, www.jama-english.jp/asia/news/vol18.pdf accessed in April 2006

Jelinek, F., Carlier, S. and Smith, J. (2004) 'AEM Validation Report and Appendices', Report EEC/SEE/2004/004, EUROCONTROL Experimental Centre, Brétigny-sur-Orge, France, www.eurocontrol.int/eec/public/standard_page/SEE_2004_report_4.html accessed in April 2006

Kar, A. and Takeuchi, K. (2004) 'Yellow Dust: An Overview of Research and Felt Needs', *Journal of Arid Environments*, vol 59, pp167–187

Lee, B.-K., Jun, N.-Y. and Lee, H. K. (2004) 'Comparison of Particulate Matter Characteristics Before, During, and After Asian Dust Events in Incheon and Ulsan, Korea', *Atmospheric Environment*, vol 38, pp1535–1545

NASA (2004) 'Severe Sandstorm in North-east China', Earth Observatory and Natural Hazards, National Agency, http://earthobservatory.nasa.gov/NaturalHazards/shownh.php3?img_id=11997 accessed in April 2006

The Nation (2005) 'Call for Massive Bus Rapid Transport System to End City's Traffic Woes', *The Nation*, Bangkok's Independent Newspaper, 13 September 2005, www.nationmultimedia.com/2005/09/13/national/index.php?news=national_18588326.html accessed in April 2006

Neil, D., Fishman, J. and Szykman, J. (2003) 'Utilization of NASA Data and Information to Support Emission Inventory Development', NARSTO Emission Inventory Workshop: Innovative Methods for Emission Inventory Development and Evaluation, Austin, TX

NRTEE (2004) 'Progress on Greenhouse Gas Emissions Trading: A Country-by-Country Review', National Round Table on the Environment and the Economy, Canada, www.nrtee-trnee.ca/eng/programs/ArchivedPrograms/Emission-Trading/overview_countries.htm accessed in April 2006

Oanh, N. T. K., Chongrak Polprasert and Nabin Upadhyay (2004) 'Improving Air Quality in Asian Developing Countries', presentation during the First Coordination Meeting of Regional Air Quality Management Programs and Initiatives, 16 June 2004, Bangkok, www.cleanairnet.org/asia/1412/articles-58466_AIRPET.ppt and www.serd.ait.ac.th/airpet/index.htm accessed in April 2006

Pramono, K. A. and Heuberger R. (2001) 'Inspection and Maintenance of Private Cars: Possible Emissions Reductions Costs and Benefits for the Car Holder', prepared for the Swisscontact Clean Air Project, Jakarta, Indonesia, www.cleanairnet.org/caiasia/1412/articles-70500_study.doc accessed in March 2006.

Sun, Y., Zhuang, G., Wang, Y., Han, L., Guo, J., Dan, M., Zhang, W., Wang, Z. and Hao, Z. (2004) 'The Air-borne Particulate Pollution in Beijing – Concentration, Composition, Distribution and Sources', *Atmospheric Environment*, vol 38, pp5991–6004

Swift, B. (2000) 'Allowance Trading and Potential Hot Spots – Good News from the Acid Rain Program', *Environment Reporter*, vol 31, pp954–959, Washington DC, The Bureau of National Affairs, Inc

Synovate (2005) 'Air Pollution Adversely Affecting 96% of Residents', Survey Findings, Synovate Global Market Research Firm, www.synovate.com/current/news/article/2005/01/air-pollution-adversely-affecting-96-of-residents.html accessed in March 2006

UIC (2004) 'Clean Coal Technologies', Briefing Paper No. 83, November 2004, Uranium Information Centre, Melbourne, www.uic.com.au/nip83.htm accessed in April 2006

UNECE (2004) 'Convention on Long-range Transboundary Air Pollution Task Force on Hemispheric Transport of Air Pollution', United Nations Economic Commission for Europe, Geneva, www.unece.org/env/tfhtap/

UNEP (2000a) 'The Atmospheric Brown Cloud: Climate and other Environmental Impacts', United Nations Environmental Programme, Nairobi, www.rrcap.unep.org/issues/air/impactstudy/index.cfm accessed in April 2006

UNEP (2000b) 'Malé Declaration on Control and Prevention of Air Pollution and its Likely Transboundary Effects for South Asia' www.rrcap.unep.org/issues/air/maledec/baseline/ and www.cleanairnet.org/asia/1412/article-69567.html accessed in April 2006

UNEP/WHO (1992) *Urban Air Pollution in Megacities of the World*, United Nations Environment Programme/World Health Organization, Oxford, Blackwell

USDOE (2002) 'Clean Coal Technology and The Presidents Clean Coal Power Initiative', US Department of Energy, Washington DC, www.fossil.energy.gov/programs/powersystems/cleancoal/index.html accessed in April 2006

USOMB (2003) *Informing Regulatory Decisions: 2003 Report to Congress on the Costs and Benefits of Federal Regulations and Unfunded Mandates on State, Local, and Tribal Entities*, Office of Management and Budget, Office of Information and Regulatory Affairs, Washington DC, www.whitehouse.gov/omb/inforeg/2003_cost-ben_final_rpt.pdf accessed in April 2006

Van Vuuren, D. P., Cofala, J., Eerens, H. E., Oostenrijk, R., Heyes, C., Klimont, Z., den Elzen, M. G. J. and Amann, M. (2006) 'Exploring the Ancillary Benefits of the Kyoto Protocol for Air Pollution in Europe', *Energy Policy*, vol 34, pp444–460

Vingarzan, R. (2004) 'A Review of Surface Ozone Background Levels and Trends', *Atmospheric Environment*, vol 38, pp3431–3442

Wang T., Poon, C. N., Kwok, Y. H. and Li, Y. S. (2003) 'Characterizing the Temporal Variability and Emission Patterns of Pollution Plumes in the Pearl River Delta of China', *Atmospheric Environment*, vol 37, pp3539–3550

Wheeler, C. (1998) 'Counting the Cost of the 1997 Haze', IDRC Report 24, August 1998, Ottawa, International Development Research Centre

WHO (2000) 'Guidelines for Air Quality', World Health Organization, WHO/SDE/OEH/00.02, http://whqlibdoc.who.int/hq/2000/WHO_SDE_OEH_00.02_pp1-104.pdf and http://whqlibdoc.who.int/hq/2000/WHO_SDE_OEH_00.02_pp105-190.pdf accessed in April 2006

WHO (2001) 'Air Management Information System (AMIS)', CD ROM 3rd edn, Geneva, World Health Organization

WHO (2003) 'Health Aspects of Air Pollution with Particulate Matter, Ozone and Nitrogen Dioxide', Report of a WHO Working Group, Bonn, 13–15 January 2003, www.who.dk/document/e79097.pdf accessed in February 2006

WHO (2004) 'Meta-analysis of Time-series Studies and Panel Studies of Particulate Matter (PM) and Ozone (O_3)', Report of a WHO Task Group, World Health Organization, WHO Regional Office for Europe, Copenhagen, www.euro.int/document/e82792 accessed in February 2006

WHO (2005) 'WHO Air Quality Guidelines Global Update 2005', Report on a Working Group meeting, Bonn, Germany, 18–20 October 2005, Copenhagen, World Health Organization, WHO Regional Office for Europe

WHO/UNEP/MARC (1996) *Air Quality Management and Assessment Capabilities in 20 Major Cities*, UNEP/DEIA/AR.96.2/WHO/EOS.95.7, Nairobi, Kenya, United Nations Environment Programme

Zhang, Y. H., Zhu, X. L., Zeng, L. and Wang, W. (2004) 'Source Apportionment of Fine-particle Pollution in Beijing', in Development, Security, and Cooperation Policy and Global Affairs, National Research Council of The National Academies, National Academy of Engineering of The National Academies 'Urbanization, Energy, and Air Pollution in China: The Challenges Ahead', Proceedings of a Symposium, Chinese Academy of Engineering, Chinese Academy of Sciences, pp139–153, National Academies Press, Washington DC, http://fermat.nap.edu/books/0309093236/html/R1.html accessed in April 2006

Annex I

AIR QUALITY MANAGEMENT CAPABILITY ASSESSMENT

Please tick "yes" ☑ or "no" ☒ to the following questions.

1. Indicators of air quality measurement capacity   25

1.1   Indicator of capacity to measure chronic health effects   3

At least one site in a residential area which has been monitoring for one year or more with a frequency of greater than one day in six for the following pollutants:

1.1.1 NO_2	Yes ☐	**0.5**	No ☐
1.1.2 SO_2	Yes ☐	**0.5**	No ☐
1.1.3 Particulate matter	Yes ☐	**0.5**	No ☐
1.1.4 CO	Yes ☐	**0.5**	No ☐
1.1.5 Pb	Yes ☐	**0.5**	No ☐
1.1.6 O_3	Yes ☐	**0.5**	No ☐

1.2   Indicator of the capacity to measure acute health effects   2.5

At least one site in a residential area which has been monitoring for one year or more and provides daily or hourly mean values, each day for the following pollutants:

1.2.1 NO_2	Yes ☐	**0.5**	No ☐
1.2.2 SO_2	Yes ☐	**0.5**	No ☐
1.2.3 Particulate matter	Yes ☐	**0.5**	No ☐
1.2.4 CO	Yes ☐	**0.5**	No ☐
1.2.5 O_3[a]	Yes ☐	**0.5**	No ☐

[a] Daily mean ozone levels are not a useful indicator since night-time levels are usually very low; therefore, daytime hourly maximum or 8-hour concentration indicators should be used for acute health effects.

There are no acute effects of lead and, therefore, no indicator.

1.3 Indicator of the capacity to measure trends in pollutant concentrations   3

At least one site in a residential area which has been monitoring for a minimum of five years capable of providing annual mean values for the following pollutants:

1.3.1 NO_2	Yes ☐	**0.5**	No ☐
1.3.2 SO_2	Yes ☐	**0.5**	No ☐
1.3.3 Particulate matter	Yes ☐	**0.5**	No ☐
1.3.4 CO	Yes ☐	**0.5**	No ☐
1.3.5 Pb	Yes ☐	**0.5**	No ☐
1.3.6 O_3[b]	Yes ☐	**0.5**	No ☐

[b] Annual mean ozone is not a useful indicator and maximum, 98th percentile, second highest value or some equivalent statistic should be used.

1.4 Indicator of the capacity to measure the spatial distribution of pollutants   3

At least three sites, one site in each of a predominantly residential, commercial and industrial area of the city, which have been monitoring for at least one year using equivalent equipment and methodologies (or those for which inter-comparisons have been conducted), with a monitoring frequency greater than one day in six, for the following pollutants:

1.4.1 NO_2	Yes ☐	**0.5**	No ☐
1.4.2 SO_2	Yes ☐	**0.5**	No ☐
1.4.3 Particulate matter	Yes ☐	**0.5**	No ☐
1.4.4 CO	Yes ☐	**0.5**	No ☐
1.4.5 Pb	Yes ☐	**0.5**	No ☐
1.4.6 O_3[c]	Yes ☐	**0.5**	No ☐

[c] The ozone sites should be located upwind and downwind in the suburbs of the city, and in the city centre, due to the secondary nature of O_3 pollution.

If mapping of pollutants had been conducted using modelling and an emissions inventory, this would be considered as meeting the indicator's criteria.

1.5   Indicator of the capacity to measure kerbside concentrations   **2.5**

A site monitoring within 3 m of the roadside or kerb operating for one year or more at least one day in six, for the following pollutants:

1.5.1 NO_2	Yes ☐	**0.5**	No ☐
1.5.2 SO_2	Yes ☐	**0.5**	No ☐
1.5.3 Particulate matter	Yes ☐	**0.5**	No ☐
1.5.4 CO	Yes ☐	**0.5**	No ☐
1.5.5 Pb	Yes ☐	**0.5**	No ☐

There is no indicator for O_3 since concentrations are very low at the roadside due to depletion by reaction with NO.

1.6   Indicators of data quality   **11**

1.6.1 Instruments calibrated at least monthly	Yes ☐	**2**	No ☐
1.6.2 Calibrations and analysis conducted using certified solutions or gases	Yes ☐	**2**	No ☐
1.6.3 Site audits conducted to compare measurements from different instruments in the network (inter-comparisons)	Yes ☐	**1**	No ☐
1.6.4 Auditing procedures conducted by an independent body	Yes ☐	**1**	No ☐
1.6.5 Sample analysis and audits performed by a laboratory with an accreditation certificate	Yes ☐	**1**	No ☐
1.6.6 Sites reviewed at least every five years to ensure they still meet the objectives of the network and hence are appropriate	Yes ☐	**1**	No ☐
1.6.7 Data are validated (critically assessed) before they are finally ratified	Yes ☐	**2**	No ☐
1.6.8 Inter-comparison exercises are conducted between different measurement techniques and/or instruments from other networks	Yes ☐	**1**	No ☐

2. Indicators of data assessment and availability   25

2.1   Indicators of the capacity to analyse data   **14**

Statistics and data analysis determined from the raw data include:

2.1.1 Means (daily, monthly, annual)	Yes ☐	**1**	No ☐

2.1.2   Maximum values (daily, monthly, annual)   Yes ☐   **1**   No ☐
2.1.3   Percentiles   Yes ☐   **1**   No ☐
2.1.4   Exceedances of national or WHO air quality
standards   Yes ☐   **2**   No ☐
2.1.5   Trends   Yes ☐   **1**   No ☐
2.1.6   Spatial distribution (mapping)   Yes ☐   **1**   No ☐
2.1.7   Exposure assessments   Yes ☐   **1**   No ☐
2.1.8   Epidemiological studies   Yes ☐   **2**   No ☐
2.1.9   Modelling with meteorological measurements   Yes ☐   **1**   No ☐
2.1.10 Prediction modelling   Yes ☐   **1**   No ☐
2.1.11 Computers are used in data assessment   Yes ☐   **2**   No ☐

2.2   Indicators of data dissemination   **11**

Air quality information about the city is available:

2.2.1   As raw data   Yes ☐   **1**   No ☐
2.2.2   In newspapers   Yes ☐   **1**   No ☐
2.2.3   On television and radio   Yes ☐   **1**   No ☐
2.2.4   On information boards in the city centre   Yes ☐   **1**   No ☐

Data are accessed through (select one):

2.2.5   Published reports which are readily available   Yes ☐   **4**   No ☐
2.2.6   Internal reports and bulletins   Yes ☐   **3**   No ☐
2.2.7   Only when requested – no formal documents available Yes ☐   **1**   No ☐
2.2.8   Air quality warnings are issued to the public during
episodes of pollution   Yes ☐   **3**   No ☐

3. Indicators of emissions estimates   **25**

3.1   Indicators of emission estimates   **8**

Estimates of emissions from the following source categories are available:

3.1.1   Domestic emissions   Yes ☐   **1**   No ☐
3.1.2   Commercial emissions   Yes ☐   **1**   No ☐
3.1.3   Power-generating facilities emissions   Yes ☐   **1**   No ☐
3.1.4   Industrial emissions   Yes ☐   **1**   No ☐
3.1.5   Cars   Yes ☐   **1**   No ☐
3.1.6   Motorcycles   Yes ☐   **1**   No ☐
3.1.7   Others, e.g. ships, aircraft   Yes ☐   **1**   No ☐
3.1.8   HGV/buses   Yes ☐   **1**   No ☐

3.2   Indicators of pollutant emission estimates   **6**

Estimates of emissions from the following pollutants are available:

3.2.1	Nitrogen oxides	Yes ☐	**1**	No ☐
3.2.2	SO_2	Yes ☐	**1**	No ☐
3.2.3	Particulate matter/smoke	Yes ☐	**1**	No ☐
3.2.4	CO	Yes ☐	**1**	No ☐
3.2.5	Pb	Yes ☐	**1**	No ☐
3.2.6	Hydrocarbons	Yes ☐	**1**	No ☐

3.3   Indicators of the accuracy of emission estimates   **9**

The inventory is calculated using (either/or):

3.3.1	Estimates based upon some actual measurements	Yes ☐	**2**	No ☐
3.3.2	Estimates based upon fuel consumption statistics and emission estimates only	Yes ☐	**1**	No ☐
3.3.3	Emissions from non-combustion processes are included	Yes ☐	**2**	No ☐
3.3.4	The inventory is cross-checked (validated)	Yes ☐	**2**	No ☐
3.3.5	Inventories are conducted at least every two years	Yes ☐	**1**	No ☐
3.3.6	Future inventories are planned	Yes ☐	**1**	No ☐

3.4   Indicators of the availability of the emissions estimates   **2**

Details of the inventory are (either/or):

3.4.1	Published in full	Yes ☐	**2**	No ☐
3.4.2	Partially available	Yes ☐	**1**	No ☐

4. Indicators of air quality management capability tools   **25**

4.1   Indicators of the capacity to assess air quality acceptability   **8**

4.1.1   Acute ambient air quality standards have been established for:

NO_2	Yes ☐	**0.5**	No ☐
SO_2	Yes ☐	**0.5**	No ☐
PM	Yes ☐	**0.5**	No ☐
O_3	Yes ☐	**0.5**	No ☐
CO	Yes ☐	**0.5**	No ☐

(Acute standards refer to those with an averaging time of one day or less)

4.1.2 Chronic ambient air quality standards have been established for:

NO$_2$	Yes ☐	**0.5**	No ☐
SO$_2$	Yes ☐	**0.5**	No ☐
PM	Yes ☐	**0.5**	No ☐
Pb	Yes ☐	**0.5**	No ☐

(Chronic standards refer to those with an averaging time longer than one day.)

4.1.3 Regulations exist to enforce compliance with air quality standards　　Yes ☐　**1.5** No ☐

4.1.4 Local air quality standards exist to take account of sensitive ecosystems　　Yes ☐　**1** No ☐

4.1.5 Air quality standards or guidelines are being introduced and/or amended in the future　　Yes ☐　**1** No ☐

4.2 Indicators of the capacity to use air quality information　　17

4.2.1 Emissions controls imposed upon:

2–3 Wheelers	Yes ☐	**1**	No ☐
Cars	Yes ☐	**1**	No ☐
HGVs/buses	Yes ☐	**1**	No ☐
Heavy industry	Yes ☐	**1**	No ☐
Light industry	Yes ☐	**1**	No ☐

4.2.2 Penalties imposed for exceeding emissions limits from:

2–3 Wheelers	Yes ☐	**1**	No ☐
Cars	Yes ☐	**1**	No ☐
HGVs/buses	Yes ☐	**1**	No ☐
Heavy industry	Yes ☐	**1**	No ☐
Light industry	Yes ☐	**1**	No ☐

4.2.3 Local air quality considered in development of new:

Roads	Yes ☐	**1**	No ☐
Industrial plant	Yes ☐	**1**	No ☐

4.2.4 Unleaded petrol available in the city　　Yes ☐　**2** No ☐

4.2.5 Additional emission controls are imposed during episodes of particularly poor air quality　　Yes ☐　**3** No ☐

Annex II

REVIEW OF QUESTIONNAIRE

The capability questionnaire has been designed to collate information on four sets of indicators (indices) which represent the key components of AQM capability:

1 Air quality management capacity index
2 Data assessment and availability index
3 Emission estimates index
4 Management enabling capabilities index.

Each indicator was allocated a total of 25 points. The overall assessment is based on an overall capability index score of 100. The total score determines the effectiveness of AQM capability on a range from minimal to excellent (see table below).

Table AII.1 *Bandings for the component and overall capability indices*

Effectiveness of capability	Component index score	Overall capability index score
Minimal	0 – ≤5	0 – ≤20
Limited	>5 – ≤10	>20 – ≤40
Moderate	>10 – ≤15	>40 – ≤60
Good	>15 – ≤20	>60 – ≤80
Excellent	>20 – ≤25	>80 – ≤100

Note: ≤ smaller than or equal to; > larger than.

The original MARC questionnaire was reviewed and revised to reflect the current situation in Asia and the increasing importance of the issues.

Air quality management capacity index

From the point of health impacts of pollutants, particulate matter (PM) is the most hazardous pollutant named in the questionnaire. However, the health effects of nitrogen dioxide (NO_2) are still less well assessed and debatable for current outdoor air concentrations. Therefore higher points were allocated to PM (0.75) and a lower one to NO_2 (0.25), the remaining pollutants had the original points of 0.5 each.

This procedure was applied for the 'indicators of capacity to measure chronic health (1.1) and acute health (1.2) effects' and to the 'indicators of acute (4.1.1) and chronic (4.1.2) ambient air quality standards', respectively. For the indicator of the capacity to measure kerbside concentrations pollutants (1.5) were given different points according to their hazard to human health: SO_2 and NO_2 (0.25), PM (0.75), CO (0.75) and lead (0.5).

Critical assessment was considered as the most important indicator of data quality within the indicators of air quality measurement capacity. Therefore, the indicator critical assessment (1.6.7) was rated with the highest points (3) reflecting the importance of data validation in data quality assessment. The next highest rankings of 2.5 each were given to monthly calibration (1.6.1) and calibration with certified solution or gases (1.6.2). The remaining indicators were allocated 1 point each.

Data assessment and availability index

The most important indicators of the capacity to analyse data within the indicators of data assessment and availability are epidemiological studies, exposure assessments, and tests of compliance with WHO guidelines or national standards. Therefore, the indicator for the usability of data for epidemiological studies (2.1.8) was rated highest (3), followed by scoring the ability to test for exceedances (2.1.4) and making exposure assessments with (2.1.7) 2 points each. The indicator 'computers are used in data assessment' (2.1.11) was not included in the modified questionnaire.

Emission estimates index

The most important 'indicators of emission estimates' for source categories are estimates for HGV/buses (3.1.8), motorcycles (3.1.6), and cars (3.1.5) while others (ships, aircraft; 3.1.7)) play only a localized role. HGV/buses and motorcycles were given 2 points each, while cars were given 1.5 points and for 'others, e.g. ships, aircraft' 0.5 points. The remaining indicators each have 1 point each.

The total score for 'indicators of emission estimates' (3.1 and 3.2) is 16 points, two points higher than in the original MARC questionnaire. This is compensated by giving a lower number of points to the 'indicators of the accuracy of emission estimates' (3.3), and attributing to the 'indicators for estimates based upon some actual measurements' (3.3.1) and 'cross-checking of the inventory' (3.3.5) 2 points each, and the remaining indicators 1 point each.

REVIEW OF THE INTERNAL LOGIC OF THE QUESTIONNAIRE

The relationship between the various indicators in the questionnaire was carefully reviewed. This review resulted in the following conclusions.

Air quality monitoring is related to the existence of air quality standards or guidelines. When no standards exist or guidelines are being applied as quasi standards or criteria for judgement, the value of monitoring is reduced. If for a certain compound there is no short-term or long-term air quality standard, air pollutant monitoring of that compound is of less value than if compliance with a standard was tested. Therefore, the following procedure was applied:

- If monitoring is performed under the 'indicators of the capacity to measure chronic (1.1) or acute (1.2) health effects', but long-term or short-term standards do not exist then the entries for the compounds in the 'indicators of capacity to measure chronic of acute health effects' were set to half the score. For example, if an indicator for acute ambient air standards (4.1.1) is 'no' for a specific compound, then, for the same compound, the indicator of the capacity to measure acute effects (1.2) is set to half its original value. If an indicator for chronic ambient air standards is 'no' for some compound, then, for the same compound, the indicator of the capacity to measure chronic effects (1.1) is set to half its original value.
- An analogous procedure is also applied for the 'indicators of the capacity to measure kerbside concentration' (1.5) if both acute (4.1.1) and chronic (4.1.2) ambient air standards do not exist.

If standards for a certain compound have been adopted but monitoring is not performed for that compound, these standards are not effective. If monitoring is performed only for some of the settings indicated in the 'indicators of air quality measurement capacity', for example not in residential areas, standards are only partially effective with respect to rational air quality management. Therefore, the following procedure was applied:

- If standards are set but no monitoring is performed, or monitoring is performed in only part of the settings, the scores in the 'indicators of the capacity to assess air quality acceptability' (4.1) should be weighted for each compound separately by the sum over the original scores over the settings, in which monitoring is performed, divided by maximum possible sum of scores, both calculated for the compound considered

■ According to the questions on emission inventories, estimates are either calculated based upon 'some' measurements (3.3.1) or on 'fuel consumption statistics and emission estimates only' (3.3.2). The meaning is that if measured emission factors are applied the estimate is better and more reliable than if only raw material input statistics and estimates (rules of thumb) are applied. Therefore, in the case that measured emission factors have been used (i.e. 'yes' in the indicator 'estimates based upon some actual measurement' (3.3.1)), the indicator 'estimates based upon fuel consumption statistics and emission estimates only' (3.3.2) should not be scored cumulatively. If the indicator 'estimates based upon some actual measurements' (3.3.1) is 1, the indicator 'estimates based upon fuel consumption statistics and emission estimates only' (3.3.2) should be set to 0. If the indicator 'estimates based upon some actual measurements' (3.3.1) is 0, the indicator 'estimates based upon fuel consumption statistics and emission estimates (3.3.2) only' may be 1 or 0, depending on the response 'yes' or 'no'.

■ If the indicator 'estimates based upon some actual measurements' (3.3.1) is 'no', the indicator 'the inventory is cross-checked' (3.3.4) should be set to 0. If the indicator 'estimates based upon some actual measurements' (3.3.1) is 'yes', the indicator 'the inventory is cross-checked' (3.3.4)' may be 1 or 0, depending on the response 'yes' or 'no'. The indicator 'emissions from non-combustion processes are included' (3.3.3) is set to the value 1 only if either the indicator 'estimates based upon some actual measurements' (3.3.1) or the indicator 'estimates based upon fuel consumption statistics and emission estimates only' (3.3.2) are not zero. The indicator 'inventories are conducted at least every two years' (3.3.5) should be zero if both the indicator 'estimates based upon some actual measurements' (3.3.1) and the indicator 'estimates based upon fuel consumption statistics and emission estimates only' (3.3.2) are zero.

The indicator 'Unleaded petrol available in the city' (4.2.4) was not included in the modified questionnaire as all cities have phased out lead in gasoline.

Glossary

Acid deposition A complex chemical and atmospheric phenomenon that occurs when emissions of sulphur and nitrogen compounds and other substances are transformed by chemical processes in the atmosphere and then deposited on earth in either wet or dry form. The wet forms, popularly called acid rain, can fall to earth as rain, snow, or fog. The dry forms are acidic gases or particulates.[1]

Acid rain Rain which is especially acidic (pH <5.2). Principal components of acid rain typically include nitric and sulphuric acid. These may be formed by the combination of nitrogen and sulphur oxides with water vapour in the atmosphere.[1]

Acute exposure One or a series of short-term exposures generally lasting less than 24 hours.[1]

Acute health effect A health effect that occurs over a relatively short period of time (e.g. minutes or hours). The term is used to describe brief exposures and effects which appear promptly after exposure.[1]

Adverse health effect Change in morphology, physiology, growth, development or life span of an organism which results in impairment of functional capacity or impairment of capacity to compensate for additional stress or increase in susceptibility to the harmful effects of other environmental influences.[2]

Aerosol Particles of solid or liquid matter that can remain suspended in air from a few minutes to many months depending on the particle size and weight.[1]

Area source An area source may be defined as a collection of similar emission units within a geographic area. Area sources collectively represent individual sources that are small and numerous, and that have not been inventoried as specific point, mobile or biogenic sources.[9] An extended pollutant emitting area such as a waste deposit or a dump is also considered as an area source.

Aerodynamic diameter The diameter of a spherical particle of the same density that, relative to a given phenomenon or property (e.g. free-falling velocity; surface area; volume; and aerodynamic properties) would behave as the particle under investigation.[3]

Agricultural burning The intentional use of fire for vegetation management in areas such as agricultural fields, orchards, rangelands and forests.[1]

Air So called 'pure' air is a mixture of gases containing about 78 per cent nitrogen; 21 per cent oxygen; less than 1 per cent of carbon dioxide, argon and other gases; and varying amounts of water vapour.[1]

Air Quality Index (AQI) A numerical index used for reporting severity of air pollution levels to the public. AQI levels range from 0 (good air quality) to 500 (hazardous air quality). The higher the index, the higher level of pollutants and the greater the likelihood of health effects.[1] In Asian countries, the AQI is the maximum value of the ratios of pollutant concentration and corresponding standard for all criteria pollutants.

Air quality standard A level of air pollutant such as a concentration or a deposition value which is adopted by a regulatory authority as enforceable. Unlike a guideline value, a number of elements in addition to the effect-based level and the averaging time must be specified in the formulation of a standard. These elements include the measurement strategy, data handling procedures, statistics used to derive, from measurements, the value compared with the standard. The numerical value of a standard may also include the permitted number of exceedances.[1]

Alternative fuels Fuels such as methanol, ethanol, natural gas and liquid petroleum gas that are cleaner burning and help to meet mobile and stationary emission standards. These fuels may be used in place of less clean fuels for powering motor vehicles.[1]

Ambient air The air occurring at a particular time and place outside of structures. Often used interchangeably with 'outdoor air'.[1]

Asthma A chronic inflammatory disorder of the lungs characterized by wheezing, breathlessness, chest tightness and cough.[1]

Atmosphere The gaseous mass or envelope of air surrounding the Earth. From ground-level up, the atmosphere is further subdivided into the troposphere, stratosphere, mesosphere and the thermosphere.[1]

Biogenic source Biological sources such as plants and animals that emit air pollutants such as volatile organic compounds. Examples of biogenic sources include animal management operations, and oak and pine tree forests.[1]

Cancer A group of diseases characterized by uncontrolled invasive growth of body cells leading to the formation of malignant tumours that tend to grow rapidly and spread.[1]

Carbon dioxide (CO_2) A colourless, odourless gas that occurs naturally in the Earth's atmosphere. Significant quantities are also emitted into the air by fossil fuel combustion.[1]

Carbon monoxide (CO)   A colourless, odourless gas resulting from the incomplete combustion of hydrocarbon fuels. CO interferes with the blood's ability to carry oxygen to the body's tissues and results in numerous adverse health effects. Over 80 per cent of the CO emitted in urban areas is contributed by motor vehicles. CO is a criteria air pollutant.[1]

Chronic exposure   Long-term exposure, usually lasting one year to a lifetime.[1]

Chronic health effect   A health effect that occurs over a relatively long period of time (e.g. months or years).[1]

Chronic obstructive pulmonary disease (COPD)   A disease process that decreases the ability of the lungs to perform ventilation. Diagnostic criteria include a history of persistent dyspnoea (an uncomfortable sensation associated with breathing) on exertion, with or without chronic cough, and less than half of normal predicted maximum breathing capacity. Diseases that cause this condition are chronic bronchitis, pulmonary emphysema, chronic asthma and chronic bronchiolitis.[4]

Dose   Once an agent enters the human body by either intake (amount of the agent that crosses the boundary) or uptake (amount of the agent that crosses the absorption boundary), it is described as a dose. Several different types of dose are relevant to exposure estimation. All these different dose measures are approximations of the target or biological effective dose (the portion of the delivered dose that reaches the site or sites of toxic action).[7]

Dose–response   The relationship between the dose of a pollutant and the response (or effect) it produces on a biological system.[1]

Dust   Solid particulate matter that can become airborne.[1]

Emission factor   For stationary sources, the relationship between the amount of pollution produced and the amount of raw material processed or burned. For mobile sources, the relationship between the amount of pollution produced and the number of vehicle miles travelled. By using the emission factor of a pollutant and specific data regarding quantities of materials used by a given source, it is possible to compute emissions for the source. This approach is used in preparing an emissions inventory.[1]

Epidemiology   The study of the occurrence and distribution of disease within a population.[1]

Exceedance   A measured level of an air pollutant higher than the national or local ambient air quality standards.[1]

Exposure   Exposure is the contact over time and space between a person and one or more biological, chemical or physical agents.[7]

Forced expiratory volume (FEV)   The volume of air that can be expired after a full inspiration. The expiration is done as quickly as possible and the volume measured at precise times; at ½, 1, 2 and 3 seconds. This provides valuable information concerning the ability to expel air from the lungs.[4]

Greenhouse effect   The warming effect of the Earth's atmosphere. Light energy from the sun which passes through the Earth's atmosphere is absorbed by the Earth's surface and re-radiated into the atmosphere as heat energy. The heat energy is then trapped by the atmosphere, creating a situation similar to that which occurs in a car with its windows rolled up. A number of scientists believe that the emission of CO_2 and other gases into the atmosphere may increase the greenhouse effect and contribute to global warming.[1]

Greenhouse gases   Atmospheric gases such as carbon dioxide, methane, chloro-fluorocarbons, nitrous oxide, ozone and water vapour that slow the passage of re-radiated heat through the Earth's atmosphere.[1]

Global warming   An increase in the temperature of the Earth's troposphere. Global warming has occurred in the past as a result of natural influences, but the term is most often used to refer to the warming predicted by computer models to occur as a result of increased emissions of greenhouse gases.[1]

Guideline   Any kind of recommendation or guidance on the protection of human beings or receptors in the environment from the adverse effects of air pollutants. As such, it is not restricted to a numerical value but might be expressed in another way, for example as exposure–response information or as a unit of risk estimate.[8]

Guideline value   A particular form of guideline. It is a numerical value expressed either as a concentration in ambient air, a tolerable intake, or as a deposition level, which is linked to an averaging time. In the case of human health, the guideline value defines a concentration below which the risk for the occurrence of adverse effects is negligibly low. It does, however, not guarantee the absolute exclusion of effects at concentrations at or below the guideline value. For odorous compounds the guideline value represents an odour threshold.[8]

Hartridge smoke unit   Percentage of opacity on a smoke filter.

Haze   A suspension in the atmosphere of extremely small (dry) particles, individually invisible to the naked eye, but which are numerous enough to give the atmosphere an appearance of opalescence together with reduced visibility.[1]

Hydrocarbons   Compounds containing various combinations of hydrogen and carbon atoms. They may be emitted into the air by natural sources (e.g. trees) and as a result of fossil and vegetative fuel combustion, fuel volatilization and solvent use. Hydrocarbons are a major contributor to smog.[1]

Indoor air pollution   Air pollutants that occur within buildings or other enclosed spaces, as opposed to those occurring in outdoor or ambient air. Some examples of indoor air pollutants are nitrogen oxides, smoke, asbestos, formaldehyde and carbon monoxide.[1]

Industrial source   Any of a large number of sources – such as manufacturing operations, oil and gas refineries, food-processing plants and energy-generating facilities – that emit substances into the atmosphere.[1]

Lead   A grey-white metal that is soft, malleable, ductile and resistant to corrosion. Sources of lead resulting in concentrations in the air include industrial sources and weathering of soils followed by fugitive dust emissions. Health effects from exposure to lead include brain and kidney damage and learning disabilities.[1]

Mobile sources   Sources of air pollution such as automobiles, motorcycles, trucks, off-road vehicles, boats and airplanes.[1]

Monitoring   The periodic or continuous sampling and analysis of air pollutants in ambient air or from individual pollution sources.[1]

Nitrogen Oxides (Oxides of Nitrogen, NO$_x$)   A general term pertaining to compounds of nitric oxide (NO), nitrogen dioxide (NO$_2$) and other oxides of nitrogen. Nitrogen oxides are typically created during combustion processes, and are major contributors to smog formation and acid deposition. NO$_2$ may result in numerous adverse health effects.[1]

Ozone   A strong smelling, pale blue, reactive toxic chemical gas consisting of three oxygen atoms. It is a product of the photochemical process involving the sun's energy and ozone precursors, such as hydrocarbons and oxides of nitrogen. Ozone exists in the upper atmosphere ozone layer (stratospheric ozone) as well as at the Earth's surface in the troposphere (ozone). Ozone in the troposphere causes numerous adverse health effects and is a criteria air pollutant. It is a major component of smog.[1]

Ozone precursors   Chemicals such as non-methane hydrocarbons and oxides of nitrogen, occurring either naturally or as a result of human activities, which contribute to the formation of ozone, a major component of smog.[1]

Particle   Small discrete mass of solid or liquid matter.

Particulate matter (PM)   Any material, except pure water, that exists in the solid or liquid state in the atmosphere. The size of particulate matter can vary from coarse, wind-blown dust particles to fine particle combustion products.[1]

Photochemical reaction   A term referring to chemical reactions brought about by the light energy of the sun. The reaction of nitrogen oxides with hydrocarbons in the presence of sunlight to form ozone is an example of a photochemical reaction.[1]

Photochemical smog   Result of reactions in the atmosphere between nitrogen oxides, organic compounds and oxidants under the influence of sunlight, leading to the formation of oxidizing compounds or possibly causing poor visibility, eye irritation or damage to material and vegetation if sufficiently concentrated.[5]

PM$_{10}$ An air pollutant consisting of small particles with an aerodynamic diameter less than or equal to a nominal 10 μm (about 1/7 the diameter of a single human hair). Their small size allows them to make their way to the air sacs deep within the lungs where they may be deposited and result in adverse health effects. PM$_{10}$ also causes visibility reduction.[1]

PM$_{2.5}$ Includes tiny particles with an aerodynamic diameter less than or equal to a nominal 2.5 μm. This fraction of particulate matter penetrates most deeply into the lungs.[1]

Pollution prevention   The use of materials, processes or practices to reduce, minimize or eliminate the creation of pollutants or wastes. It includes practices that reduce the use of toxic or hazardous materials, energy, water and/or other resources.[1]

Polycyclic aromatic hydrocarbons (PAHs)   Organic compounds which include only carbon and hydrogen with a fused ring structure containing at least two benzene (six-sided) rings. PAHs may also contain additional fused rings that are not six-sided. The combustion of organic substances is a common source of atmospheric PAHs.[1]

Regional haze   The haze produced by a multitude of sources and activities which emit fine particles and their precursors across a broad geographic area.

Residence time   The average time a molecule or aerosol spends in the atmosphere after it is released or generated there. For compounds with well-defined sources and emission rates, this is estimated by the ratio of the average global concentration of a substance to its production rate on a global scale. It is a function not only of the emission rates but also of the loss rates by chemical and physical removal processes.[6]

Smog   A combination of smoke and other particulates, ozone, hydrocarbons, nitrogen oxides and other chemically reactive compounds which, under certain conditions of weather and sunlight, may result in a murky brown haze that causes adverse health effects. The primary source of smog in California is motor vehicles.[1]

Smoke   A form of air pollution consisting primarily of particulate matter (i.e. particles released by combustion). Other components of smoke include gaseous air pollutants such as hydrocarbons, oxides of nitrogen and carbon monoxide. Sources of smoke may include fossil fuel combustion, agricultural burning and other combustion processes.[1]

Source Any place or object from which air pollutants are released. Sources that are fixed in space are stationary sources and sources that move are mobile sources.[1]

Stationary source Non-mobile sources such as power plants, refineries and manufacturing facilities which emit air pollutants.[1]

Stakeholders Citizens, environmentalists, businesses and government representatives that have a stake in or concern about how air quality is managed.[1]

Sulphur dioxide (SO_2) A strong smelling, colourless gas that is formed by the combustion of fossil fuels. Power plants, which may use coal or oil high in sulphur content, can be major sources of SO_2. SO_2 and other sulphur oxides contribute to the problem of acid deposition.[1]

TSP Total suspended particulate (TSP) is particles of solid or liquid matter such as soot, dust, aerosols, fumes and mist up to approximately 30 μm in size.[1]

Volatile organic compounds (VOCs) Carbon-containing compounds that evaporate into the air (with a few exceptions). VOCs contribute to the formation of smog and/or may themselves be toxic. VOCs often have an odour, and some examples include gasoline, alcohol and the solvents used in paints.[1]

NOTES

1   California Air Resources Board, www.arb.ca.gov/html/gloss.htm accessed in April 2005

2   WHO (1994) *Assessing Human Health Risks of Chemicals: Derivation of Guidance Values for Health-based Exposure Limits*, Environmental Health Criteria 170, Geneva, World Health Organization

3   Willeke, K. and Baron, P. A. (1993) *Aerosol Measurement: Principles, Techniques, and Applications*, New York, Van Nostrand Reinhold

4   CMD (1997) *Taber's Cyclopedic Medical Dictionary*, 18th edn, Philadelphia, FA Davis Company

5   ISO (1994) 'International Standard 4225: Air quality – General aspects – Vocabulary', Geneva, International Organization for Standardization

6   IUPAC (1997) *Compendium of Chemical Terminology – IUPAC Recommendations*, compiled by A. D. McNaught and A. Wilson, International Union of Pure and Applied Chemistry, Oxford, Blackwell Science Ltd

7   WHO (2000a) *Human Exposure Assessment*, Environmental Health Criteria 214, World Health Organization, Geneva, www.inchem.org/documents/ehc/ehc/ehc214.htm#SectionNumber:1.5 accessed in August 2006

8   WHO (2000b) 'Guidelines for Air Quality', WHO/SDE/OEH/00.02, World Health Organization, Geneva, website as on p175

9   US EPA (2001) 'Introduction to Area Source Emission Inventory Development', Emission Inventory Improvement Program, Document Series, vol 3, Chapter 1, Eastern Research Group, Inc., Area Sources Committee, United States Environmental Protection Agency, www.epa.gov/ttn/chief/eiip/techreport/volume03/iii01_apr2001.pdf accessed in August 2006

Index

Page numbers in *italics* denote references to Figures and Tables.